Introduction to
Scientific Geographic Research

FOURTH EDITION

Introduction to
Scientific Geographic Research

.

The Late L. Lloyd Haring
Formerly of Arizona State University

John F. Lounsbury
Arizona State University

John W. Frazier
State University of New York—Binghamton

FOREWORD BY HARM DE BLIJ

WCB McGraw-Hill

Boston, Massachusetts   Burr Ridge, Illinois   Dubuque, Iowa
Madison, Wisconsin   New York, New York   San Francisco, California   St. Louis, Missouri

WCB/McGraw-Hill

A Division of The McGraw-Hill Companies

Book Team

Editor *Jeffrey L. Hahn*
Developmental Editor *Lynne M. Meyers*
Production Coordinator *Audrey Reiter*

President *G. Franklin Lewis*
Vice President, Publisher *George Wm. Bergquist*
Vice President, Operations and Production *Beverly Kolz*
National Sales Manager *Virginia S. Moffat*
Group Sales Manager *Vincent R. Di Blasi*
Vice President, Editor in Chief *Edward G. Jaffe*
Marketing Manager *John W. Calhoun*
Advertising Manager *Amy Schmitz*
Managing Editor, Production *Colleen A. Yonda*
Manager of Visuals and Design *Faye M. Schilling*
Production Editorial Manager *Julie A. Kennedy*
Production Editorial Manager *Anne Fuerste*
Publishing Services Manager *Karen J. Slaght*

President and Chief Executive Officer *Mark C. Falb*
Chairman of the Board *Wm. C. Brown*

Cover design by Tara Bazata

Cover image by ©Comstock, Inc.

CONTENTS

LIST OF FIGURES

LIST OF TABLES

FOREWORD

During my days as a graduate student in the Department of Geography at Northwestern University in the 1950s, the curriculum leading to a Ph.D. included several hurdles all of us approached with trepidation. There was the field camp, held during the summer at what is now the University of Wisconsin, Platteville, a six-week geographic bootcamp that taught us techniques ranging from plane-tabling to crop identification, and from soil analysis to interview methods. There was the Master's-level oral examination—that feared, penultimate test of our knowledge in theory, substance, and methodology. (I was lucky. My examiners got into a prolonged argument among themselves over something I said, and when this abated, my time was up.) There was the dissertation proposal—an exercise best begun at least a year before its final draft was due, or so the veterans warned us.

But nothing we feared was comparable to that menacing brainchild of what we regarded as an overzealous faculty, the research paper, otherwise known as the "Pressure Problem." At a certain time during the final year in residence, each of us was required to produce an original report based on field research. This exercise was called the Pressure Problem because it had to be carried out—from the opening of the envelope containing the desideratum to the presentation of the paper, complete with finished maps and photographs—in exactly two weeks. Some doctoral candidates never did manage it; the Pressure Problem could be a nemesis.

I was reminded of this challenge when reading *Introduction to Scientific Geographic Research*. What a difference this volume would have made to those of us who, having opened the infamous envelope, had no idea how to go about solving the posed problem! Here, in just 150 advice-packed pages, authors Lounsbury and Frazier guide students, in clear and efficient language, through the intricacies of problem definition, research design, data generation, survey research, analytical methods, and the options arising from modern developments in automation. Students are counseled on ways to approach a geographic research problem, how to use models, and how to write the research report when the results are in. There are pointers on citations, illustrations, revisions, and abstracts, and recommendations on the use of statistical methods and cartographic representations. In short, this book is a mine of information for the student confronted, perhaps for the first time, with the challenge of original geographic research. Despite the intricacies of their subject, the authors manage to maintain an accessible, engaging tone.

Since 1989, when I began discussing geography on national television on a regular basis, I have received many hundreds of letters. One of the questions most often asked is a variant of the theme "How do you *do* geography?" Here is the book to answer it.

H. J. DE BLIJ

PREFACE

Similar to the previous three editions, the fourth edition is designed to serve as a text or guide for college students experiencing their first involvement in an actual research project. It is likely that most college students majoring in geography will first be exposed to geographic research methods at the advanced undergraduate level or at the beginning graduate student level. Most geography departments normally offer either formal courses concerned with introductory research methods and techniques or informal courses such as seminars, tutorials, or guided study courses. At the present time, there is little published material designed specifically to serve as a guide for students engaged in their first research project. Often they are frustrated in their attempts to organize their work and make order out of chaos. They find it difficult to define the research problem and associated hypotheses, collect meaningful data, employ the proper methods and techniques of analysis, and reach logical conclusions.

Modern geography is a discipline with many recognized and well-established subfields. Many of these subfields have, over the course of time, developed highly specialized research techniques utilizing sophisticated data-collecting and data-processing equipment. It is not possible nor is it the intention of the authors to provide a book that includes the advanced research techniques and tools applicable to all the subfields within the discipline of geography. Rather, the major purpose of the book is to serve as a guide to enable the student to develop an orderly, scientific outlook and to provide a framework of reference to direct the student's enthusiasm and efforts along productive tracks.

The book is written specifically for students who have little or no previous research experience. Over the course of three editions, it has been demonstrated that it may be used in any course concerned with introductory research methods as well as an aid in preparing term papers or projects in a variety of other geography courses. The book has been designed purposely to be general rather than specific, broad rather than restricted, and flexible rather than rigid. The instructor, therefore, can select what is useful to his or her specific needs as well as add substantive material without difficulty.

In October 1982, Professor L. Lloyd Haring died in a traffic accident. Professor John W. Frazier, Harpur College of the State University of New York, Binghamton and John F. Lounsbury, Arizona State University have joined forces and have coauthored the fourth edition. The authors have taught introductory research methods for many years at several colleges and universities, at both the undergraduate and graduate levels. The authors received their graduate training at different universities, and each has his own areas of specialization. It is hoped that this diversity in the authors' training, interests, viewpoints, and professional specialties is reflected in the book's broad approach, pertinent to the overall discipline of geography, rather than an emphasis on specific subfields.

The fourth edition follows the basic format and design of the three earlier editions, which proved to be useful and successful guides to introductory research. However, the fourth edition represents a significant expansion reflecting new technological developments in the discipline. Each chapter has been revised, updated, and expanded. New chapters have been added to include computer mapping, geographic information systems, geographic surveys, and geographic report writing.

CHAPTER ONE

The Nature of Scientific Research

Since the earliest of times, the human race has had an inherent curiosity about the environments and events that take place in the world in which they live. They have sought answers to a host of perplexing problems and questions they have encountered. Some of the answers they have accepted are theological in nature, formal and authoritative, and not to be questioned. Others are aesthetic or emotional, and inclined to be informal based on an individual's preferences and opinions. Still others are the result of a person's experience and overall common sense, and often take the form of an old proverb or adage. All of these types of answers to the questions or problems posed are not, as a rule, subject to vigorous testing, and there has been a great variation in answers from one culture to another, as well as from time to time. The scientific method, however, provides a framework within which the answers to questions or problems do not vary from place to place or from time to time. Anywhere on the face of the earth, and the moon as well, two plus two equal four, and this is an undisputable fact today as well as anytime in the past or the future. The scientific method as we know it today, in the context of human history, is very new. It is estimated that over 90 percent of all scientists who have ever lived are alive at the present time.

"Beauty is truth" said John Keats, and the artist, the artisan, the scholar, and indeed every person of intelligence and goodwill has valued truth both for itself and because it is only through truth that we gain knowledge. There is no single group who has the key to truth or the path to its door. Each group makes its contribution in its own way. One way to truth is through the social and physical sciences to which geography belongs. This is the way you have chosen, and whether your selected field is in physical geography, social geography, or a combination of the two, you have decided to make your contribution to truth and knowledge primarily as a scientist.

The scientist is fortunate to live and work in the present time because it is an accepted part of the twentieth century thought that science and its methods are of social value. Although there are other paths to human satisfaction (and by this is meant satisfying the needs and desires of society), the one which we follow is paved with the discoveries and recorded contributions of scientific research. Whether the desire is to go to the moon, to cure the sick, to build a better bridge, or to teach a child to read, we turn to science for the solution. This is sufficient justification for science and it is the reason that social and physical sciences are important and respected segments of higher learning today.

The Framework of Science

Science seeks the truth in an objective, rational manner through a process of controlled inquiry. The guiding impulse of science is curiosity, and the beginning point is a question for which an answer is sought. Throughout this search, all terms used are defined precisely, and all procedures are described carefully. All processes of deductive or inductive reasoning are clearly demonstrable, and each tool or technique employed is explained. Scientific conclusions are never expected to be accepted on faith or solely on the authority of the researcher. Research in science is not for everyone. It is only for those who are willing to demonstrate the validity of each step, from the posing of the question to its eventual solution. It is for those who not only want to know, but who will not be satisfied with an answer unless the process and reasoning behind it can be verified by reason, analysis, and rigorous testing.

The methods we identify today with science began to flourish in western Europe near the turn of the seventeenth century. While many scholars contributed to the system, it was an Englishman, Sir Francis Bacon (1561–1626) who received the most credit for first describing the method we now refer to as the Scientific Method. He set forth its tenets in his historic work Novum Organum (or The New Method).[1] Prior to this time answers to questions had been accepted primarily by unquestioned belief in authority, either based on faith or the conjecture of logic. Bacon expressed the opinion that these methods of problem solving were often wrong. He espoused a process that began with a problem for which a solution or "hypothesis" would be suggested. This hypothesis, assumption, or educated guess then guided the research through observation, analysis, synthesis, and finally, to the conclusion. The conclusion, then, would come only after the hypothesis was thoroughly tested. On the basis of the evidence gathered, it was accepted or rejected as an answer to the original question. During this process, the scientist remained objective, suspended judgment until all evidence was gathered, and recorded his data carefully and precisely.

Scientists today follow procedures very much like those set forth by Bacon. Scientific researchers' answers are always objective, verifiable, and impartial, and whenever possible, they have mathematical precision. They are not content to deal in vague terms, such as "more" or "less," but seek to discover exactly how much more or how much less. They weigh, measure, and calculate systematically, each step logically following the one before. The final answer is based on the preponderance of evidence carefully analyzed and is as correct as the facts available at a given time will allow.

Methods of Research

Scientific researchers select the procedure that is best for solving the problem with which they are concerned. In general, there are three major "methods" of procedure recognized: the experimental, the normative, and the historical.

The experimental method is used mainly by the physical scientists but it may be used in geography. To use it, it is essential that one is able to control the variables that affect the dependent variable, thereby determining the result. By rigorous control of all variables employed in the experiment, the physical scientist is able to describe, analyze, and predict with a high degree of precision. Such experimental work has become the epitome of science for many people.

An example of experimental research would be an experiment in which botanists were interested in knowing the exact amount of light, moisture, and heat that produced the most rapid plant growth. The researcher might place a number of plants in separate containers in which they could control the three independent variables they were concerned with in the study. By varying each one separately, and keeping careful records of the effect on plant growth in each container, a generalization could be made concerning the factors and their quantities that contribute to optimal plant growth.

---

1. James Spedding, Robert Leslie Ellis, and Douglas Denon Heath, eds., *The Works of Francis Bacon* (London: Longmans & Co., 1879), vol. 1, pp. 70-117.

In those areas of research where variable control is not possible or feasible, the normative method is a widely used method of investigation. The essence of this method is to observe events and evaluate the observed processes with a view of establishing constant relationships or norms. In this method, description and analysis are as important as they are in the experimental method. Where relationships are discovered, prediction becomes possible. Frequently, geographic norms take the form of a dependent variable "Y" varying proportionately with an independent variable "X." From such a relationship, laws based upon these norms may be formulated, and their predictable variation may be stated.

Since it is difficult, and often impossible, to control the independent variables affecting the behavior of people, the normative method is usually the one most applicable to problems in social geography. Thus, for example, the effect of language, religion, or income on voting patterns may be determined by observing past elections and measuring the variations in the number of votes as it varies in relation to varying amounts of the influencing factors. In physical geography an example might be in the problem determining the rate at which temperatures decrease with rising elevations. By taking repeated recordings as one ascends and descends a mountain or while traveling in an airplane, an average or normal rate of change can be established. Thus a normative generalization is possible.

A third method of research is one in which the researcher neither controls nor evaluates the variables observed. Here the major objective is to record observations accurately. The historical method relies heavily on source detection, evaluation, and analysis of findings in determining solutions or conclusions to the research problem. An excellent example of this method is found in medical research where a relationship between death and stimulating social events has been established. Between 1875 and 1915, the death rate in Budapest, Hungary, dropped markedly before the observance of Yom Kippur. From this and corroborating information, the evidence suggests that dying is a form of social behavior.[2]

While widely used by historians, this method is often combined with the normative method in historical geography studies. By this means, the researcher may describe historical events and establish past spatial distributions and areal associations.

---

2. David Phillips, "The Vital Buoyancy of Optimism," *Time*, September 5, 1969, pp. 58–60.

Geography as a Research Discipline

Since the dawn of time people have wanted to know about places, both the immediate environment and those that could not be directly experienced. This natural curiosity accounts for much of the popularity of the subject since Strabo (64 B.C.–20 A.D.) wrote Geographia 2,000 years ago. In Medieval Europe the travels of Marco Polo received instant fame, and in more recent times Immanuel Kant (1724–1804) assigned to geography the task of studying all of those associations that are associated in space. To him this study of space meant the study of places and the interaction of everything of significance within that space. Many credit him with being the father of modern geography because he emphasized the study of interaction among phenomena, but essentially he remained committed to the importance of place. Geographers today continue this emphasis, whether they concentrate on the relation between people and environments or choose to study relations among social or physical phenomena only. Because humans do not possess an innate knowledge of area, it is only through education (formal and informal) that such knowledge is gained.

Geography is the branch of academic research and study primarily concerned with the acquisition of spatial knowledge within a scientific and highly structured framework. It is the major discipline that is concerned with the identification, analysis, and interpretation of spatial distributions of phenomena and their areal associations as they occur on the surface of the earth.

The field of geography as a whole does not specialize in one particular set of phenomena,[3] but on relations between and among phenomena. Like science itself, the discipline focuses on method. All phenomena that occupy space are grist for geographic analysis. The goal of all science is to describe, analyze, and finally, to predict. Geography does these things and is capable of doing them very well. Geography is a science because its research techniques use the scientific method and produce results similar to other scientific disciplines.

Geography is a vigorous science concentrating on the concept of space and spatial relations which, along with time and composition of matter, comprise the three major parameters of concern for all of science:

Of the three great parameters of concern to scientists, space, time, and composition of matter, geography is concerned with two. Geography treats the man-environment system primarily from the point of view of space in time. It seeks to explain how the subsystems of the physical environment are organized on the earth's surface, and how man distributes himself over

3. Jan Broek, *Geography: Its Scope and Spirit* (Columbus, Ohio: Charles E. Merrill Publishing Co., 1965), p. 5.

the earth in his space relation to physical features and to other men. Geography's organizing concept, for which "spatial distributions and space relations" are a verbal shorthand, is a triscaler space. The scales comprise extent, density, and succession. Geography's theoretical framework is developed from this basic concept.[4]

There is reason to believe that geography is perhaps the most logical discipline for understanding the systems of life. Geographic research investigates and tests hypotheses advanced to explain problems of spatial associations. It is essentially concerned with place, location, territory, distance, space, and their interactions. When the methods of science are applied to these problems, the results are worthy of interest to the entire scientific community.

There will always be workers in scientific disciplines who wish their areas of specialization were not a science, for science is a stern taskmaster. For those who wish to rely on opinions, who want instant answers, or who prefer conjecture or faith as a method of gaining answers to problems, science is not a suitable approach. The goal of scientific research is not a collection of interesting observations about some place, event, or process. Such studies may be artistic, interesting, fun to do, and even useful; but they do not fall within the realm of scientific geographic research.

Whether it be a social or a physical study, geographic research is the application of the scientific method toward solving spatial questions. As such, it must have a problem, an hypothesis, accurate data, a presentation of findings, and a conclusion. It may describe or analyze or predict; or all three purposes may be served in one study. But it must follow the procedures of science if it is to be considered a scientific research project.

The Scope and Subfields of Geography

Geography is a broad discipline concerned with the spatial distributions of phenomena and their causal associations. Geography examines both the structure of distributions and the processes that create them. The discipline consists of many well-established subfields. As a discipline geography is definable by its persistence in dealing with a set of disciplinary "questions." These "questions" have frequently been described as our "themes," or the ties that bind the field of inquiry together. Pattison provided one very useful description of geography's traditions nearly three decades ago. He suggested that any piece of geographic research could logically fall into one of the following themes: environ-

4. Ad Hoc Committee of Geography, Earth Sciences Division, *The Science of Geography* (Washington, D.C.: National Academy of Sciences-National Research Council, 1965), p. 1.

mental, human-environmental, area study, or spatial.[5] The environmental theme consumes all geographic research focusing on environmental processes and patterns of the biosphere, atmosphere, and hydrosphere, while the human-environmental theme encompasses research dealing with the linkages between people and their environments, including the impact of humans in the use (or abuse) of environments, such as wilderness. Area study, or regional analysis, is one of geography's oldest themes and involves the definition of an area (region) using boundaries and an analysis of phenomena and their interrelationships that make such an area distinct. The fourth theme, spatial, often is interrelated with the first three themes, but is distinct by its focus on spatial relationships: the relationship between elements in a distribution (nodes, linkages, etc.) and the causal processes. An example of the spatial theme is the use of diffusion theory to explain the spread of a phenomenon through space over time. The theory stipulates the nature of the pattern and process for specific phenomena using generic explanations, or hypotheses.

Geographers have debated Pattison's thematic framework and have offered some modifications. Any such framework, however, clearly demonstrates geography's disciplinary core that has centuries-old roots and poses questions about our earthly environment(s) and humankind's relationship with it (them). These questions, of course, are broad. As a result, as the discipline evolved, the number of subfields increased, but the *themes* remained constant. Today, within geography's broad themes, the Association of American Geographers recognizes over fifty topical or systematic subfields and over sixty major areal or regional specializations (see Table 1.1).

In addition, the Association has established Specialty Groups to provide a mechanism for geographers with similar research interests to meet, communicate, and exchange ideas. At the present time, there are thirty-nine Specialty Groups (see Table 1.2). The number increases or decreases as research interests change over time.

Dualism in Geography

It is often a puzzle to non-geographers that our discipline can study so many different kinds of things, and indeed it has caused some serious thought on the part of geographers. As a consequence we commonly divide our subject into Physical Geography and Human Geography. This dichotomy is focused on the type of phenomena studied. The human geographer studies such things as economic, social, political

---

5. See William D. Pattison, "The Four Traditions of Geography," *The Journal of Geography*, Vol. 63 (1964), pp. 211-216.

TABLE 1.1

Major subfields of geography

Topical Proficiencies	Areal Proficiencies
01 Administration	01 World
02 Agricultural Geography	06 North Polar Region
03 Applied Geography	07 South Polar Region
04 Arid Regions	08 North America
05 Audio-Visual Materials and Techniques	09 Anglo-America
06 Biogeography	10 Canada
07 Cartography, General	11 Territories
08 Climatology	12 British Columbia
09 Coastal Geography	13 Prairie Provinces
10 Cultural Ecology	14 Ontario
11 Cultural Geography	15 Quebec
12 Economic Development	16 Atlantic Provinces
13 Economic Geography	17 USA (AAG Division Boundaries)
14 Educational Geography	18 New England
15 Energy	19 Middle States (NY, NJ, DE, E. PA)
16 Environmental Perception	20 Middle Atlantic (MD, DC Metro Area)
17 Environmental Studies (Conservation)	21 Southeastern
18 Field Methods	22 Southwestern
19 Gender	23 Pacific Coast
20 Geographic Information Systems	24 Great Plains/Rocky Mountains
21 Geomorphology	25 West Lakes
22 Historical Geography	26 East Lakes
23 History of Geography	28 Latin America (Hispanic America)
24 Land Use	29 Middle America (Caribbean)
25 Librarianship, Geographical	30 Mexico
26 Location Theory	31 Central America
27 Manufacturing Geography	32 West Indies and Bermuda
28 Marine Resources	33 South America
29 Marketing Geography	34 Northern South America and West Coast
30 Medical Geography	35 Brazil
31 Military Geography	36 Southern South America
32 Natural Hazards	37 Europe
33 Oceanography	38 Scandinavia
34 Physical Geography	39 Western Europe
35 Planning, Regional	40 British Isles
36 Planning, Urban	41 Southern Europe
37 Political Geography	42 Central Europe
38 Population Geography	43 Eastern Europe
39 Quantitative Methods	44 USSR
40 Recreational Geography	45 Asia
41 Regional Geography	46 Korea
42 Remote Sensing	47 China
43 Resource Geography	48 Japan
44 Rural Geography	49 Southeast Asia
45 Social Geography	50 South Asia

46	Soils Geography	53	Southwest Asia (Mideast)
47	Teaching Techniques	54	Africa
48	Transportation and Communication	55	North Africa
49	Tropical Geography	56	West Africa
50	Urban Geography	57	Central Africa
51	Water Resources	58	East Africa
		59	Africa South of Congo
		60	Australia
		61	New Zealand
		62	Pacific Islands

Major areas of specialization or subfields of geography as recognized by the Association of American Geographers.

TABLE 1.2

Specialty groups

Africa	Geomorphology
Aging and the Aged	Hazards
American Indians	Historical
Applied	Industrial
Asian	Latin America
Bible	Mathematical Models and Quantitative Methods
Biogeography	Medical
Canadian Studies	Microcomputers
Cartography	Political
China	Population
Climate	Recreation, Tourism, and Sport
Coastal and Marine	Regional Development and Planning
Contemporary Agriculture and Rural Land Use	Remote Sensing
Cultural	Rural Development
Cultural Ecology	Socialist Geography
Energy and Environment	Soviet and Eastern European
Environmental Perception and Behavioral Geography	Transportation
Geographic Information Systems	Urban
Geography Perspectives on Women	Water Resources
Geography in Higher Education	

Specialty Groups of the Association of American Geographers as of July, 1990. New Specialty Groups are established as new areas of research develop.

or cultural things. Thus, a study in marketing geography or transportation would be an example of economic geography. The social geographer might be concerned with where crime occurs, the distribution of religion, or the movement of aged people. Political geography concentrates on nations, boundaries, voting patterns, and similar political things. The cultural geographer studies the development of different cultures, and is closely related to the historical geographer in that they both concentrate on things or events that have occurred in the past but affect the distribution of things in the present.

Physical geography is commonly divided into fields concerned with the lithosphere, biosphere, hydrosphere, and atmosphere. Examples would be a study of landforms, plant distributions or wild animal life, oceanography, or the atmospheric distribution and behavior of climates. Of course these are only a few examples of the many studies that develop from each of the major disciplines.

Another dichotomy in the field of geography has developed out of the approach the geographer takes to the subject. If he focuses on one area and considers a number of significant things as they interact within the area, it is considered to be regional geography. Regional geography has a long tradition, and for a period during and prior to World War II it dominated the discipline in America. While it has lost some of its research interest in recent decades it remains an essential part of geography and is often cited as being the epitome of the study of place.

The contrasting study for Regional Geography is Topical Geography, sometimes called Systematic Geography. This branch of the subject concentrates on the one type of phenomena, such as plant life, and studies it wherever it occurs. Such studies do not have to cover the entire earth; for example, a topical study might involve only the animal production in the Corn Belt of America. In such studies it may be seen that regional and topical studies often overlap. The Corn Belt is a region, but is filled with many topical subjects of research. In fact the two approaches often merge, especially in such areal studies as those of two cities. The classification of such studies depends essentially on whether the intent is on understanding the area as an entity or understanding the interrelationship of a few categories of phenomena. Essentially, the differences between regional and topical geography are differences in approach rather than differences of subject matter. They often merge or overlap in many areas of research.

Still another recently conceived dichotomy (applied versus theoretical) revolves around the basic goals of a given research project or projects. *Applied Geography* normally deals with practical problems that are meaningful at the present time. For a variety of reasons, a precise definition of applied geography that is completely compatible with the thinking of all geographers does not exist. However, if applied geography refers to the application of geographic theory and research methodologies toward the solution of current problems facing an area, region, country, or the world, such a broad interpretation would be acceptable in principle to most geographers. *Theoretical Geography* on the other hand, is primarily concerned with the search for knowledge and truth, regardless of any practical value or applications. This does not mean, however, that theory and practice are truly dichotomous. Applied geography, as a science, shares the goals of theoretical geography, but as an applied science, it is an extension of purely theoretical geography. Applied geography, like other applied sciences, has several attributes: it is user-oriented, action-oriented, and extends the scientific method to include evaluation and implementation stages to achieve the user- and action-orientations.

Examples of theoretical geography might be the study of climatic conditions during the Pleistocene Ice Age or the development of a mathematical model to explain the diffusion of an innovation, such as a new product and its adoption (purchase) at a specific time (period) in a given area (region). Although the goal of theoretical geography is not practical significance per se, the results of theoretical geographic research may, at one time or another, have practical applications. For example, the theoretical factors that predict why and when a person purchases a new product are of obvious value to a marketer or business person, even though the study was not necessarily designed for that person or the business's particular product. An applied geography research design would add two stages: evaluation of the research findings for specific goal and strategy formulation, and implementation of change by controlling outcomes. In this sense, it is creating future geographics by using research results to guide policy decisions about places—where to market a new product first, choosing which site for a new shopping center, where to locate a new landfill, selecting a new route, or predicting environmental impacts for a specific area based on a specific plan of action. Thus, applied geographic research is primarily concerned with providing answers to a present problem, but, on occasion, may point to new avenues for theoretical research. Applied geography has experienced a rapid growth in recent years.

Changing Research Patterns and Areas

In most all fields of science and often in short periods of time, new avenues of research develop. Emphasis on a given type of research increases rapidly also. Generally, the reasons for new research emphases or thrusts are threefold. First, a new idea or concept is developed. For the biologist, evolution is an example, and, most recently, DNA and genetic behavior. For the physicist the atom, and later quarks and the nature of matter. Second, the development of new technological instruments such as the electron microscope, radar, and a host of others have provided new data and, as a result, encouraged new research efforts. Finally, changes in the national or world economy or trends in geopolitics have fostered new research directions. The launching of Sputnik led to the creation of the National Science Foundation which in turn funded a great number of research projects and programs, particularly in the physical and engineering sciences. Depressions, civil rights movements, and world economic conditions opened the doors to new research in the political, social, and economic fields. Geography has been greatly influenced by new ideas such as the Central Place Theory and others. Also influential are new technological developments such as aerial photo and remote sensing imagery and electronic computers, to name but a few. Ups and downs of the national economy and social structure has spurred research in social problems such as housing, medical needs, racial problems, population movements, and urban and regional planning.

The research environment in any field is dynamic, and changing emphases and thrusts develop constantly and often quickly. Overall, the field of geography is sensitive to changes of technology, new ideas, and national and global conditions. Although it is difficult to predict the precise nature of geographic research thrusts in the future, it is a safe assumption to postulate that the profile of research topics will be different in the foreseeable future than the present time.

Steps in Geographic Research

Geographic research usually begins with a spatial distribution the geographer must explain. It is the primary task of the geographer to account for why something is located in the place it is found. In arriving at this end the researcher usually starts with the locative position of a set of phenomena (things) that are capable of being mapped, although they do not need to be physically observable. For example, mountains may be observed and mapped but voting patterns or religions cannot be seen, although they may be mapped, and thus display a spatial distribution suitable for describing and explaining. In research, such a distribution is designated as the dependent variable if it is the purpose of the researcher to account for its locative situation by associating it with other variables whose position help explain the original distribution with which we are concerned. For example, the location of tamerisk in western United States has a definite pattern that may be partially explained by the distribution of the independent variable of stream pattern. The geographer is trained to look for and find such related distributions, and to explain the connection among the variables that are interrelated in place. For that reason the field of geography is often defined as the study of areal association or as areal relations. The two terms mean essentially the same thing, although the second implies a greater degree of cause and effect. Of course, such cause and effect may be present even though it is not the specific purpose of a particular study to prove such a relationship.

After the researcher has selected a subject and identified the phenomena to be investigated, the success of the project depends on completing a series of specific tasks. It is important that tasks be done in a proper order to avoid confusion; for as each task is completed, it determines the specific nature of other steps immediately following. The amount of time and effort expended on any given step varies depending upon the nature of the research problem. In general, these steps in the order that they should be worked upon may be defined as follows:

1. *Formulation of the research problem,* or the asking of a previously unanswered question in exact terms. This also includes the precise determination of the areal extent of the matrix within which the research work will be conducted. A research problem may be concerned with a micro-area, such as a city block; or it may be concerned with a macro-area as large as a continent.

2. *Definition of hypotheses,* or formulating a theory, assumption, or a set of assumptions that are yet unproven but are accepted tentatively as a basis for investigation.

3. *Determination of the type of data to be collected* pertinent to the research problem. The specific nature of the research problem and the size and areal extent of the matrix will determine the type of data that must be obtained, the way the data will be classified, how the data will be collected, and whether or not the matrix will be surveyed in its entirety, or sampling procedures will be employed.

4. *Collection of data,* which involves the analysis of published materials, the use of field techniques, or perhaps the deployment of data-collecting instruments.

5. *Analyzing and processing the collected data,* which involves the selection of appropriate cartographic and statistical methods of analysis. Further, this final step includes stating the conclusions and determining if the proposed hypotheses are confirmed or denied.

In the chapters to follow, these procedures will be examined and discussed in detail.

Defining Geographic Problems

.

The exact nature of a specific geographic research design may depend upon the final stage of the project—the writing and presentation of the research work to a given and, sometimes select, audience. Geographic research is written, presented, and, perhaps, published in several ways. The major categories are the thesis or dissertation to be submitted to the academicians; articles or papers for professional journals that will be read by professional people; technical reports that describe some aspect or all of the research project's methods and results; semi-professional articles that may have some interest to the non-professional as well as the professional; proposals submitted to foundations or academicians for approval or for general information purposes; and articles or chapters for textbooks. These various types of writing will be discussed in a later chapter, but the point for the moment is that the framework of the overall research project will be determined by the nature of the group or population for whom the research is being done. These potential audiences vary greatly, and the writing and presentation of the research results must be designed for their purposes. This, to a large degree, will determine the overall research design as such.

All research, however, has four basic elements: (1) there must be a problem or a question to be answered, (2) data and information bearing on the question must be collected, (3) the collected data must be compiled and analyzed in some relevant fashion, and (4) a conclusion must be stated—Do the results of the research throw any light on the question asked? What does it all mean? These four elements are the "bare bones" of research design. Often there are other connecting steps between bridging the basic elements.

The first step in any research plan is the careful selection and identification of the problem. Problems usually arise as the result of a feeling of concern or the need for more precise knowledge. Wherever the student notes a lack of knowledge in the corpus of geography, a potential problem exists. A proper research problem is a situation of concern in geography that is described, bounded, and focused in order that concentrated study may be applied to it. For example, while studying urban distributions, a geographer might notice a lack of locational information concerning cities with populations between 100,000 and 200,000. If the student feels that additional information is needed before an understanding of urban distributions can be acquired, *a problem* has been experienced. Since problems are actually unsolved questions, one way to define the specific research problem is to ask a question such as "What are the relationships between urban sprawl and internal population mobility in cities with populations between 100,000 to 200,000?" Assuming that the results of the study are meaningful, the research has contributed some knowledge that will be of value in the understanding of the overall question of the areal dynamics of medium-sized cities.

The number of "good" research problems in geography is infinite. Many times, such problems are defined by scholars in books, dissertations, or professional papers. In fact, it is a somewhat standard practice to conclude a study with suggestions for further research. The serious student of geography will analyze published materials concerning the subject, as well as materials developed by related disciplines. From the subject matter or the principles presented in the literature, the student may identify gaps that appear to justify additional research. There is no better way to formulate a problem than for the researchers themselves to sense its need and state it in their own words. In doing this, the first step is to select specific phenomena whose locational presence in an area is of concern, then proceed to explain their presence and spatial arrangement. The phenomena to be studied may be physical or cultural, concrete or abstract, objects or ideas. They may be house types, drainage patterns, votes, or religious concepts. Any material or nonmaterial thing that can be identified, classified, and located is proper subject matter for geographic study. The geographic problem contains the element of what, where, and why—a distribution of phenomena (what) whose location or spatial characteristics (where) are to be explained by their association, or lack of association, with other phenomena (why).

The mere existence of an unanswered question does not necessarily assure the basis of a suitable research problem. There may be inherent difficulties concerning a problem or difficulties that will raise serious doubts as to its research feasibility. Is it of interest to the researcher? Probably it will be if it has been discovered and developed by the researcher's own efforts. But if it is not interesting to the individual researcher, it is doubtful that the research should be undertaken. This is especially true in theses and dissertations where considerable time and effort will be required. The quality of work will probably suffer because of the tedium of the research work; and the lack of interest may be related to other factors that will seriously affect the product.

One possibility that could detract from the interest of a problem is its apparent lack of value for a given individual. If the student does not have some curiosity about a problem, that problem probably does not have sufficient motivating value for the student to expend the sustained research effort necessary for its solution. The main object of a research project is for the apprentice research worker to learn to do research and to demonstrate this acquired skill. The results of the research need not be of practical value, nor need they make a significant contribution to the discipline. They must, however, be of some importance to the student, even if only to satisfy an intellectual curiosity.

Beginning research students run a special hazard in this respect. Quite often, their uncertainty makes them especially susceptible to suggestion. While a conscientious advisor is careful not to pressure the advisee into an unsuitable topic, such a situation can occur. Students are well advised to have a topic of interest to them in mind when they discuss the matter with an established researcher. In this regard, the late Professor Good describes a situation which is not uncommon. The beginning research student went into Professor Good's office, and this was the ensuing confrontation:

> "I've got to write a Master's thesis," says he, "and I'd like to talk to you about a topic." The statement ends with a slight upward inflection as if, in spite of its grammatical form, a sort of question were implied. After an awkward pause Mr. Blank (the student) repeats that he would like to talk about a thesis topic. Whereupon the Editor (and Professor) suggests that he go ahead and do so.
>
> It transpires, however, that the Editor-Professor has misconceived Mr. Blank's meaning. He has no topic to talk about. In fact, instead of coming with a topic, he has come to get one. He looks so expectant, too; purely, as one might say, in a receptive mood. . . . He gives the impression of having just learned about this thesis business, and of being entirely open-minded on the subject.[1]

---

1. H. G. Good, "The Editor Turns Professor," *Education Research Bulletin*, VI (September 14, 1927), pp. 252-53.

This account, while containing an element of humor, occurs too frequently to be completely humorous. If the research project is to be a pleasant exercise, as it should be, the problem should be the writer's choice and should reflect a personal interest. The student should profit from the advisor's suggestions, but should not depend upon the advisor to define a problem or to outline the study.

After it is decided that a problem exists and that a solution to the problem will be of value, the research student should determine whether a solution is possible or even probable. For instance, one might think of very interesting problems, such as determining the effect of the lost continent of Atlantis on past climatic conditions in Europe. However, the possibility of testing any hypothesis set forth as a solution for this problem is so remote that it is unacceptable for research work. Serious reservations should be entertained if the problem is of such a nature that a solution is not likely, or if the qualifications of the researcher are not adequate to solve the problem. In some cases, the lack of qualifications may be corrected by the acquisition of new skills or by mastery of additional research tools. Even experienced scholars find it beneficial at times to return to the classroom in order to learn skills necessary for the solving of new problems.

A common weakness with many interesting problems is that sufficient existing data are lacking, or new data cannot be obtained for a successful solution. This is the weakness of the hypothetical problem concerning Atlantis; and many contemporary problems have the same inherent weakness. While it may be possible to obtain the needed data by extensive field study, beginning students should be aware of the magnitude of the task. They should make a careful inventory of their resources—time, money, knowledge—and if the problem cannot be solved within their limits, they must either redefine and reduce the scope of the problem or implement their resources.

In the last analysis, a satisfactory research problem in geography is one which is of interest to the researcher, one in which sufficient data concerning it may be obtained, and which focuses on areal associations and spatial relationships.

In every stage of problem development and articulation, it is imperative that the student read widely concerning the subject. Any literature in geography or related fields that touches upon the problem should be reviewed (see Appendix A). The reasons for this are obvious, as the researcher must know what type of data is available, or may be obtained, and whether or not the problem has already been investigated. A satisfactory solution to a problem may present new data; it may develop a new methodology; or it may bring existing data together in a new way to make it more meaningful. Whatever the primary focus of the problem, the researcher will not wish to invest valuable time solving one that has been solved previously. Nor would the researcher desire to fail in a solution because the problem could not be solved. Squaring a circle or inventing the wheel may seem to present problems, but one is impossible and the other unnecessary!

Determining the Matrix (Study Area)

An important aspect of formulating the research problem is determining precisely the areal extent of the research area or matrix. There is no standard or generally accepted size of a matrix; significant research may be accomplished in very small or very large areas. However, the size of the matrix will bear directly on the types of phenomena that can be studied, as well as on the scale, detail, and classification of the data to be collected.

For example, a student researcher might become interested in changing land use patterns in the United States and pose the question "What is the relationship between types of land use and population densities in the United States?" If the problem is so stated, it would be in order to illustrate the spatial distributions of land use and population. It is possible to construct such illustrations for the entire country but only by using very general categories of land use, such as urban land, agricultural land, forest land, grassland, and the like, and by using broad categories of population densities. However, at this scale, the resulting maps would be so highly generalized that no definite conclusions could be drawn for specific areas.

Consider the situation in which the student is still concerned with the same general problem but reduces the study area to one county. It is now possible to classify land use data in more explicit terms; and what was urban land may now be subdivided into residential, commercial, industrial, and so forth. What was agricultural land may be divided into cropped land (even the type of crops), pasture land, fallow land, and the like. In the same manner, the categories of population densities may be made more detailed. This information shown spatially results in maps that are more refined and accurate than are those for the country as a whole, and the relationships become clearer.

From the results of this hypothetical study, the student may want to pursue the matter further, desiring to know how commercial land use in the central business district of the city and commercial land use in outlying shopping centers are related to population densities. The matrix may now be reduced to a few square blocks in size. This makes it possible to obtain the actual square footage of each commercial use, as well as the exact population, block by block. The researcher may find that some commercial uses are always located close to dense populations, while some types of commercial uses may be relatively far removed from residential areas.

The matrix is the framework within which the research is conducted and which determines the detail of the data collected. Usually, the smaller the matrix, the more specific will be the data; and the larger the matrix, the more general will be the data and the likelihood that sampling procedures will be employed.

In determining the matrix, the delimitation criteria used and the rationale for selecting this particular research area over all others should be stated clearly. For example, if a research problem is concerned with some aspect of wheat farming in the northern Great Plains, the first questions that will be asked are, where are the northern Great Plains, and what delimits them? why not study the entire Great Plains area? or, why not study only the southeastern quadrant of the northern Great Plains. Often, the matrix of study may be justified on the basis that superficial evidence indicates that a given phenomenon or activity appears to be concentrated in the proposed area more so than in others, or that there exists a widespread feeling of concern about a particular situation in the proposed area, or perhaps that there is a need for new information about the proposed area.

The boundaries of the research area may often coincide with a physical feature, such as a valley or a river basin, or a political division or other cultural boundary. The matrix need not always be a contiguous area, because comparing two or more areas is not uncommon. Also, the definition of the boundaries of the matrix may be influenced because essential existing data are available in one area but not in another. For example, one county may have accumulated information that is highly significant to the research problem, but this information is lacking in adjoining counties or areas.

Stating the Problem

Once a problem and its matrix are judged suitable for research, the problem should be stated carefully. At this point, it is essential and natural that a positive, confident frame of mind exist. The student has determined that the problem is pertinent, interesting, and that it can in all probability be solved within the limits of his resources. This is the problem, and the researcher is now ready to state in exact terms what he or she intends to do. This statement is not only important to the researcher but also informs others precisely what will and will not be done. The success or failure of the research project may depend upon the proper statement of the problem.

There is no one way in which a problem must be presented, but its statement should always be clear and concise. There should be no doubt about the need for this research nor about the subfield of the discipline that will profit from its solution. A question is one good way to pose a problem. Frequently when one wants an answer, a question is asked. Such questions are to the point and afford pegs on which answers may be hung. If the problem is presented as a statement, it can be changed in form to be a question. For example, the geographer may present a problem concerning the relationship between cash grain farming and flat land as a statement, or the investigator might pose the question, "What are the relationships between cash grain farming and low slope land in the American Corn Belt?" Since any problem is by its nature an unanswered question, this latter form is an accepted scientific method of posing it.

There are some important aspects to be noted in the example just given. In the first place, a complete statement of problem must define the terms it uses. Other persons may have no way of knowing what is meant by cash grain farming, low slope land, or the American Corn Belt. Until these things are known, and in terms that can be tested for possible solutions, it cannot be determined if a problem exists. Thus, additional questions must be asked, such as, what is low slope land? what are the parameters of slope? what is cash grain farming? what is the areal extent of the American Corn Belt? and, what tests will be used to determine if a relationship exists? In answer, a cash grain farm might be defined as one on which 50 percent or more of all farm income is from the sale of grain. The American Corn Belt might be defined as the area within which 20 percent of the cultivated land is devoted to corn production any given year. Low slope land might be defined as having slopes of 1° to 5°.

This procedure of defining terms is referred to as "operationalizing" the problem. Until the terms used are described in such a way that all can understand what is meant, there is no way in which a testing operation can be performed on any hypothesized relationship. In the following section, the formulation of suitable hypotheses will be discussed. For now, it is sufficient to keep in mind that the use of words or terms not susceptible to an operational definition usually remain nebulous, if not meaningless. "If no operation can be performed, it is highly doubtful that two human minds can get close enough to the subject to discuss it intelligently."[2] Now that it is known what is meant by low slope land, cash grain farming, and the American Corn Belt, the researcher can proceed to establish the type and degree of relationship existing between the two variables with the defined matrix.

Many problems in geography are questions concerning the spatial relationship of two or more variables. Normally, the question to be answered must be stated in such a way that the tentative answer, the hypothesis, may be tested to determine if it will be accepted or rejected—or with what reservations it may be accepted. The formulation of the problem, then, is wedded in its inception to the hypothesis. The scientific research problem exists in order that a scientific research solution may be sought. If the research problem is clearly stated and the research area well defined, the remaining steps may be time-consuming; but the researcher can rest assured that his subsequent efforts will be organized and that they will be expended along with most productive lines.

Nature of the Hypothesis

It is possible that some research problems may not have a formally stated hypothesis. If a research study is concerned with a problem for which little or no information exists, it may not be possible to formulate any reasonable assumptions or hypotheses. The formulation of hypotheses implies some knowledge of the problem and the research area. For example, if research is proposed to determine the spatial distribution of permafrost depths in Greenland, or the cultural traits of primitive peoples in the Amazon Basin, there may not be sufficient existing data upon which to base meaningful hypotheses. The answers to basic questions such as these would contribute, in themselves, to the overall pool of knowledge and would be accepted as worthwhile research.

In other cases, the problem statement or the nature of the problem includes a well-defined question or purpose and a separate formally stated hypothesis is redundant. The hypothesis is helpful if it further defines the problem. The statement of the problem is the identification of a felt need to know. The hypothesis is a reasonable way to meet that need. It is a proposition that is assumed to offer a possible and reasonable solution to the problem. Together the problem and hypothesis guide the research investigation.

Certain characteristics may be observed in a good hypothesis. First, it is capable of being expressed as a question. For example, the hypothesis that temperature change is associated with elevation may be expressed as the question, "Is temperature change associated with elevation?" Second, the hypothesis may be stated in a negative way: "Temperature change is not associated with elevation." This type of hypothesis statement is a null hypothesis and is useful in certain problems. For example, let us suppose that a new illness developed in a community that the medical researchers designated as "Illness A." It is important to ascertain the cause of the malady quickly, and several research teams begin work. One team was given the problem in which a null hypothesis stated "bacteria strain B4 is not the cause of Illness

A." Let us assume that the research was successful and proved without a doubt that bacteria B4 had no connection to Illness A. The cause of Illness A is still unknown, but the bacteria in question can be eliminated as a possible cause. Subsequent research by this team and other research teams can now focus their efforts on other possibilities and eliminate them one by one until the culprit is identified. The nature of the problem and problem statement determines whether a hypothesis or null hypothesis is best suited for the subsequent research to be undertaken.

While our purpose is not to include discussion of statistical analysis of data (see Chapter 6), students should be aware that hypotheses often are subjected to statistical analyses for verification. The *null hypothesis* in such cases is the primary hypothesis. It is denoted as H_0. The burden of "proof" is on its rejection. An *alternative hypothesis*, H_1 is any hypothesis that is different from the null and is typically set up to parallel it. If the statistical test leads to a rejection of the null hypothesis, the alternative is "accepted." Such tests require clearly stated alternatives and the selection of the appropriate statistical tests, including criteria that describe the degree of probability of the result. Here our concern is not with the statistical options, but with the notion of stating the null and alternative hypotheses. A few examples are useful.

Typically, the null hypothesis is stated in a precise manner:

H_0: There has been no increase in real per capita income in the United States since 1980.

The alternative is often less precise and involves a range, or even generalized expectation, if the factor under study had some effect. In short, the null hypothesis is often stated in a way contrary to the researcher's belief; the alternative is stated in opposition to the null:

H_1: There has been some gain in real per capita income in the United States since 1980.

L. J. King demonstrated the utility of the null hypothesis in statistical testing more than twenty years ago:

> Consider a simple hypothesis to the effect that the average distance traveled by farm families to purchase groceries in a region is 5 miles. Assume it is known on the basis of past studies that these distances have a variance of 4 miles, in which case the standard deviation is 2 miles. A random sample of 36 farmers is to be selected for the purpose of testing the hypothesis above. . . .
>
> In a formal statistical sense, the hypothesis which is tested in such situations is that there is no significant difference between the true mean

value for the universe (the average distance traveled for all farmers in the regions) and the hypothesized value, in this case 5.0. This hypothesis of no difference is the *null hypotheses, H_0*. The alternative hypothesis H_2, in this example is that the true mean is either greater or less than 5.0.[3]

Third, and finally, a hypothesis should be capable of being answered with a "yes," a "no," or a "maybe." In formal terms, it is capable of being accepted, rejected, or not rejected. The difference between these hypotheses and everyday questions all people ask themselves about observed phenomena is one of degree, and not of kind. As stated in Chapter 1, the scientific method is no more than sound reflective thought applied to a specific question.

In research work, the hypothesis is a statement that might be called an educated guess, an informal hunch, an assumption, a suggestion, a supposition, or a conjecture. Whatever it is called, its usefulness depends upon how well it condenses an array of facts into a statement that can be investigated and tested. It is this investigating and testing that elevates it to the status of a hypothesis. The better the researchers understand the facts related to the problem, the more educated their "guesses" will be. Hypotheses must be based on a sound rationale that clearly illustrates the connection among the dependent and independent variables. Such relationships must be based on logic, findings from other studies, field observations or some combination of sound reasons cogently stated. The more soundly based the rationale, the more likely it is that the researcher will arrive at a satisfactory conclusion.

If one were attempting to chart the shortest route by automobile to New York City from Los Angeles, one would be faced with a large number of possible alternatives. Many could be eliminated as being too improbable to consider (or "test," in research language); but a myriad of routes and subroutes would still remain. Perhaps someone who has just returned to Los Angeles from New York claims that the shortest way to New York is through Denver. This may be a reasonable supposition, and untrained people accept such informed guesses and suppositions as truth. However, until it is tested against other probable ways, the Denver route cannot be accepted as the shortest way. After it is measured against a possible shorter route, it may then be rejected or accepted as the shorter of the two routes. But it cannot be accepted as the shortest route on the basis of comparison with only one other route. As other reasonable possibilities are eliminated, there will be a point at which the hypothesis may legitimately be accepted as correct. The reason for the hypothesis becomes more apparent when one attempts to solve a problem without one.

---

3. L. J. King, *Statistical Analysis in Geography* (Englewood Cliffs, NJ: Prentice-Hall, 1969), p. 72.

The hypothesis is not always stated as a clear and separate corollary of the research problem. When it is not, the detection of the precise purpose of the research may be more difficult. For example, a problem may be stated in a manner that includes the hypothesis: "The purpose of this paper is to determine if Republican voters in New England tend to be located in rural areas." The problem here would be accounting for the locational pattern of the Republican voters in New England, and the hypothesis is that this distribution is associated with "rural" conditions (precisely defined). A slight modification sometimes encountered is the making of the statement in the form of an hypothesis: "It is hypothesized that rural conditions account for the locational pattern of Republican voters in New England." In both cases, it would seem that a more easily understood method of informing the reader of the exact nature of the problem and the hypothesis would be to state the problem clearly as a question, and then, in a separate statement, advance the hypothesized answer.

Determination of Hypotheses

Usually, there are two ways of arriving at tenable hypotheses in geographic studies. One is to consider the distribution of the phenomena the researcher wishes to explain and to seek reasonable explanations in the laws that control that phenomena. The other is to consider the pattern, location, and density of the phenomena and to compare this distribution with that of other supposedly related data. Each of these methods warrants individual consideration.

In geographic research, the starting point is the distribution of some phenomena a researcher wishes to analyze in terms of where they are and why. This is called a dependent variable since its locational pattern depends on some other variables (phenomena) that effect it. For example, in explaining the location of steel mills, two related variables might logically be iron and coal sources. The first method, then, of arriving at logical hypotheses is to turn to the laws that govern the areal location of the dependent phenomena and to determine how much of the activity should be expected at a certain place. By knowing the proportions of iron, coal, and other materials in a ton of steel, and through a process of evaluating this information in relation to transportation costs and markets, a reasonable hypothesis could be formulated. It would probably take the form of "At point X, a steel mill would be expected." In this case, the hypothesis could be tested by observing how closely the theoretical location corresponds with actual location.

It is obvious that such hypotheses require considerable knowledge or information before an acceptable solution to the problem of the location of a dependent variable can be expected. Because of the complicated nature of this information and the need for precision, such hypotheses are often stated mathematically with the use of numbers and symbols. Such statements are called *mathematical models.* Just as a table model of a house may express accurately the dimensions of someone's home, so, too, a small conceptual model which is stated mathematically may express a real life situation. For example, the amount of travel expected between City 1 and City 2 might be predicted by considering the populations of the two cities and the distance between them. This is done by the "interaction model" which is stated as $i = \dfrac{P_1\,P_2}{d}$ where "P" is the population of City 1 and City 2, and "d" is the distance between them. The interaction model, the simplest form expressed here, has been tested and refined; until today its many forms have a certain degree of accuracy in explaining actual transportation patterns.

Geographic hypotheses are often suggested as a result of the geographer's comparing the distributional patterns of a dependent variable with other known distributions. In this method, the dependent variable is usually mapped, and then this map is visually compared with other maps. On the basis of this comparison, an "educated guess" is made concerning the association between two of the maps. This guess is then studied in order to determine if an association does indeed exist, and if so, to what degree. As a rule, the locational arrangement of a variable under consideration is affected by many other variables acting upon it. The researcher must assemble a sufficient number of these independent variables that might explain the pattern of a phenomenon he proposes to study. For example, if the problem is "Why is the distribution of crime in Century City arranged in this particular pattern?" the possible hypotheses may include the distribution of: (1) schools, (2) income, (3) broken homes, (4) poor housing, (5) ethnic groups, and an infinite number of other variables. The researcher cannot hope to explain 100 percent of the distribution of such crime. Instead, he must attempt to select the significant variables effecting its location and settle for the degree of explanation desired in his problem design.

This method of hypothesis determination was used in the research on yellow fever in Panama. Maps of various possible related phenomena were compared with a map showing the areal extent of yellow fever until a likeness was detected between the fever map and a map of the distribution of the anopheles mosquito. A relationship was hypothesized, tested, and accepted. After elimination of the mosquito, yellow fever was brought under control, and the completion of the Panama Canal thus became possible. A more recent, but similar, situation occurred in a study of jaw cancer in Africa. In this case, a chance remark that the cancer was not found in southern Africa led to the problem: "Why does the tumor occur on the northern side of a given line across Africa and not on the other side?" Comparative map analysis detected an association between the line of cancer and that of rainfall, temperature, and eventually, a virus-carrying insect. Thus developed one of the major discoveries in cancer research in our lifetime.[4]

As the body of geographic literature develops, more laws governing geographic relationships may be expected. From such relationships will come the source of future hypotheses. However, for the foreseeable future, many of the hypothetical solutions will be obtained from maps, ideas gathered from reading and observations, commonsense deductions, and similar sources. These hypotheses will be tested by measuring the extent to which they express actual locational patterns. The tested and accepted hypotheses, after being subjected to extensive reexamination to confirm their validity, will become the laws of geography. Eventually, researchers will be able to turn to these laws for new hypotheses. With each tested and confirmed hypothesis, geographers will become capable of describing and analyzing the locational pattern of a specific phenomenon and of predicting its spatial relationship with other areally associated phenomena.

---

4. Bernard Glemser, "The Great Tumor Safari," *Today's Health,* XLVI (September, 1968) p. 44 ff.

CHAPTER THREE

Formulation of the Research Design

.

After the student has defined the problem, matrix, and related hypotheses, it is necessary to reflect on the entire project and then design an operating plan. This "work plan" might be compared to the blueprint a construction engineer requires before beginning to build a new structure. It controls the phases of inquiry so that procedure decisions are made before a situation arises. The research design includes not only the blueprint of the project to be undertaken, but also the materials needed, the tools required, the cost involved, and the time schedule of anticipated progress.

There are many reasons for designing such a plan. In situations where an advisor, a committee, or supervising professor are involved, presentation of the plan may be required before the proposed project is approved. If outside funds are to be sought, the funding agency will request such a plan well in advance of the proposed time schedule. Certain information may be unavailable if the research purpose and proposed data are not fully and clearly explained.

The most important reason for this design, or work plan, is that it provides the framework to allow the researcher to organize time and resources. Only the most hard (and foolish) person would embark on an important task without having a goal, a plan to reach it, and a deadline time to be there. Houses have been built without blueprints, but no modern, competent builder would attempt such a project. No builder will proceed without deciding what parts are to be built first and at what approximate time each stage should be completed. In the building industry, it is common practice to post a bond stipulating the date on which a contracted structure will be completed. Financial ruin would rapidly overtake a contractor who worked without a blueprint and a timetable. Inefficient use of time and efforts is the result of research without a work plan.

The most important parts of a work plan are an orderly arrangement of the steps involved, a time schedule, and a list of equipment necessary for the completion of the steps involved. Also, the finished work plan may be presented to others who are involved or interested in the study, and it becomes a guide for the entire project. As the study progresses, specific parts of the work plan may be expanded, altered, or eliminated as new information becomes available.

The Title

The beginning of the work plan is the selection of the title of the research study. Although this may be a tentative title, subject to modification after the paper is complete, the most appropriate title for the proposed research should be selected at this time. It should be short, but not so brief as to obscure the field of study and the essence of the research. The title should indicate the subfield of geography concerned, the research matrix, the time period, and the purpose of the study. A title such as "Crop Harvesting and Migrant Labor in California: 1940–1960" fulfills the requirements of indicating space, time, subfield, and purpose. It also tells others at a glance the major variables to be studied.

Purpose and Problem

After the title, the work plan should contain a brief but pertinent description of the area of geography in which the proposed study falls. This description should include the need for the information anticipated in the study and the specific gaps in the literature that will be filled. It may seem difficult to envision results before the study is begun, but the formation of tentative conclusions is an important and often a most neglected step in research planning.

In this part of the proposal, major terms to be used are defined in such a way that other persons will know what the terms mean and how they will be used. If the study is of village population change, the reader must know the size of a village and know how much population increase or decrease is necessary before a "change" is indicated.

Whenever possible, the anticipated direction of the study should be stated simply. If, for example, the problem is to determine the effect of "X on Y," this should be stated. The researcher may be interested in the effect of rainfall (X) on corn yields (Y), or income (X) on Republican votes (Y), or transporting costs (X) on gasoline prices (Y), or any other phenomena that may be identified and measured. Let us assume that a study is focusing on the variation of the maximum temperature at a number of locations in the researcher's home state on a particular day. Further, let us assume that the researcher has hypothesized elevation as the factor associated with temperature variation. Readers can be clearly and simply informed of the proposed problem in the following manner: "In this problem, there is one dependent variable (T) and one independent variable (E). The relationship between the variables will be clarified by answering the following questions, 'To what extent are the variations in T explained by the variations in E?' "

If more than one independent variable is hypothesized as affecting the variation of temperature (elevation or E and precipitation or P, for example), the hypothesized relationship is stated in the same manner: "To what extent is the variation of T related to the variation of P? Do the combined E and P variables provide a significant explanation of T variation?" Given the operational definitions of the variables (T, E, P), there is no doubt about what the researcher intends to investigate.

Survey of the Literature

Although not always essential, it usually is necessary that the researcher include a preliminary survey of the literature in the work plan. Such a survey is made in order to weigh the feasibility of making the study. It clarifies one's thinking to write a statement describing and evaluating the available literature pertaining to the proposed study. A second value of such a survey is to give the advisor some information on which to base guiding opinions and suggestions. In describing the literature, three points should be emphasized:

1. The amount of work that has been done previously on the subject,
2. An indication about the strong and weak areas of the existing literature, and
3. The trends pertinent to the research problem as revealed in the survey of the literature.

Procedure, Tools, and Time Schedule

The major portion of the work plan is a three-step sequence of planning: (1) the tasks considered necessary to complete the problem, (2) the tools required, and (3) the estimated time required for their completion. The first step is to make a list of individual tasks that must be completed. After this list is drafted, the individual items can be rearranged, combined, and divided into a reasonable sequence.

In the arrangement of the sequence of tasks, the order in which they are performed is not necessarily the order in which they are presented in the completed report. A map of the study area may be on page one of the introductory chapter in the finished study, but it may be scheduled for completion with other maps after the final draft has been written. As research proceeds, rearranging may be in order; but it is necessary at this stage that a tentative plan be envisioned.

After the succession of steps has been decided upon, the steps and the particular tools needed for each one should be reviewed. The tools may be *physical,* such as a particular drafting tool for a specialized map, or they may be *conceptual,* such as the knowledge for designing a stratified random sample. At this point, the researcher will be able to make an inventory of available resources and note which items are available and which ones must be obtained. Again comparing the geographer to a builder, it is at this point that an inventory of materials and needed skills is made so that they may be ordered and be available at the time they are needed.

The third step in this sequence is to check each item and to estimate the time required for each step. It is also advisable here to make note of any expenses that will be incurred in addition to the time investment of the researcher. In the event that either time or cost exceeds the researcher's resources, some arrangement must be made. This may involve modification, such as the elimination of a questionnaire for a personal survey. In some cases, it is also possible to trade time for costs. If time is crucial, some part of the work (perhaps typing of the completed paper or drafting maps) may be accomplished by employing help. It is well to consult with the advisor if there is any question involving the legitimate work of the researcher and the work that may be done by others. No researcher should claim credit for work done by others.

During this phase of the planning, some rearranging may become necessary. Perhaps a book that is needed in Step 1 must be ordered, and Step 2 can be shifted to Step 1 while the student is awaiting its arrival. A step may not be possible until a certain skill is acquired. If a course in sampling is needed or if field skills are required, the researcher may wish to take a course before completing his or her survey.

A question might arise at this point about how the researcher can estimate the time required to finish a specific task. When it is not known, the best estimate must be made. For example, the book on order probably will take two weeks to arrive. The course in sampling will end on a certain date. The amount of time needed to make the map may be determined by comparing it with other maps made. The time to read the book can be estimated from the researcher's own reading speed and from the number of pages in the book.

It is not expected that time estimates will be perfect. Nevertheless, it is necessary to determine the general time requirements for various tasks and to formulate an overall timetable, or the researcher is courting disaster. Deadlines should be made in a reasonable manner, and it is good to allow for about 10 percent more time than anticipated. Without a time schedule, the "law" of Parkinson takes effect (the time needed to complete a task expands to occupy the time available for its completion). Having a deadline puts the researcher on the initiative and alerts the mind to prepare for the task ahead. It makes organic thinking possible, an essential requisite important in research.

There are other points that should be kept in mind concerning a time schedule. Some tasks are easier and some more difficult than envisioned. If one step is completed ahead of schedule, the researcher should go on to the next. This is also related to the variation in the productivity of the researcher. One may have an extremely active day and wish to postpone a relaxation period in order to take advantage of this productivity. Related to this is the fact that all individuals have particularly productive times of the day. If it is really necessary to work thirty minutes cleaning the typewriter and locating paper, these tasks should be done during some period of little creativity.

Time Planning

Before assigning exact periods and dates of completion to individual tasks, the researcher should establish general categories of time to be invested in separate phases of the research. There are no hard rules established for this allocation of time, but some general guidelines will help the beginning student. As more experience is acquired, modifications to this schedule can be made that better suit the individual's work tempo.

Most research work has two points rather firmly established—a starting date and a completion date. The starting date is usually immediate, and the completion date is determined by the college or university schedule, funding agencies, and other outside forces. Within these limits, the researcher sets the beginning and ending dates to best fit personal needs. The length of this time period has some influence on scheduling; but as a rule of thumb, the work is performed in two stages. Perhaps in two-thirds of the time available, the researcher is involved with preparing the study, arranging, gathering, and analyzing data. Specific items include formulating the problem and hypotheses, selecting the title, surveying the literature, preparing the outline, making the tentative bibliography, collecting data, analyzing and testing data, and organizing notes under appropriate groupings.

The second phase, the remainder of the time, is assigned to the work of actually preparing the information found in the research for presentation in a written report. In order of time these steps are:

1. A logical analysis of the subject matter preparatory to arranging it in the proper sequence for presentation,
2. The outline for the completed report,
3. The writing of the first rough draft of the report,
4. The rewriting and polishing of the rough draft,
5. The assembly of the final supportive materials (maps, tables, bibliography, etc.),
6. The writing of the final draft,
7. Selection of a final title, dedication statements, index tables, abstracts, and other related activities necessary for final presentation of the study.

It is during the completion phase of the report that the final draft is typed. Some time must be allowed for it, whether someone is employed to do the actual typing or not. Also, there may be a need during the final stage to review Step 1 in order to fill gaps in the study, ones that became apparent during the logical analysis of material and the completion of the final outline.

After the major segments of the research have been assigned time limits, specific items may be assigned. The detail desired will vary greatly depending upon the type of study, the length of time involved, and the individual characteristics of the researcher. One would probably not wish to list the day that each book is to be read. If a certain book is to be secured through interlibrary loan, however, the researcher must anticipate when the book is needed so he or she can place an order sufficiently in advance and complete its use by the time the book must be returned. In this regard, it is definitely advisable to plan the points at which significant works should be obtained and completed. It is also advisable that the researcher keep a diary of this daily progress. Much research has been accomplished with little detail in planning; but the researcher who organizes work within the structure of the total project will be able to meet critical deadlines, and the final product will reflect this careful planning.

Each research study is unique, and time requirements for each task will vary. However, assuming that the research project is scheduled to

TABLE 3.1

Sample Work/Time Master Schedule

Task	Proposed Time
Phase I. Planning Actual Research	12 weeks (total)
A. Survey of Pertinent Literature (Identification of significant professional papers, reports)	1 week
B. Acquisition of Pertinent Source Materials (base maps, statistical reports, surveys, etc.)	1 week
C. Collection of Data (library and field research)	6 weeks
D. Analysis and Processing of Data (compilation, evaluation, selection of data)	4 weeks
Phase II. Preparing the Written Report	5 weeks (total)
A. Preparation of Report Outline (outline of chapter and major headings)	1/2 week
B. Preparation of First Draft (compilation and first writing of complete report)	1 1/2 weeks
C. Revision of First Draft (rearranging, editing, modifying, rewriting)	1 week
D. Assembly and Preparation of Final Supportive Materials (drafting maps, illustrations, tables; assembly of bibliography)	1 week
E. Preparation of Final Draft	1 week

be completed within the temporal framework of a college semester (approximately seventeen weeks in duration), the student researcher might construct a work/time master schedule based on the best possible estimates (see Table 3.1).

It is evident that Table 3.1 is highly generalized and that it should not be considered as inflexible or rigid. Rather, it should be looked upon as a sample or guide that indicates the general magnitude of the various tasks comprising a research study, a guide the student should consider in constructing a master schedule to fit existing requirements.

In the event that the research study represents a major piece of work, such as a Master's thesis, it is most likely that a longer period of time should be allocated to writing the report (Phase II) than is shown in Table 3.1. Often, research projects take longer than one semester, but the percentage of time devoted to various tasks most likely will remain about the same.

Research Design Issues: A Summary of Important Questions

We have discussed the importance of a clear definition of the research problem and of clearly stated hypotheses. In this chapter we have presented important aspects of planning an actual research project. This is a good time to reflect on important issues related to these planning elements and to raise other issues associated with any research design. The latter provides a framework for the remainder of this book. These issues can be stated in the form of questions and are presented in Table 3.2. The questions that appear under topics A through D involve issues addressed in Chapters 1 through 3, while those under E and F are issues that guide remaining chapters of this book.

Up to now, one of the things that we have emphasized is the importance of clearly stating the research problem. It is also important to place the research topic within its proper perspective (context) and to clearly articulate the purpose of the proposed research, as shown in Table 3.2.

TABLE 3.2

Questions Addressing Adequacy of the Research Design[1]

A.   *Research Problem/Topic of the Proposed Geographic Research*
1.   What is your research *topic*?
2.   Is the research topic/problem *clearly stated*?
3.   Is the topic (research) placed within its proper *context* or *perspective*?
4.   Is the *purpose* of your proposed geographic research clearly articulated?

B.   *Literature Review*
1.   Did you read *sufficient literature* on your research topic to understand the relevant theory(ies)?
2.   Is your literature summary adequate in *scope, detail,* and *clarity*?
3.   Are both variables and findings of previous research clearly explained?

C.   *Hypotheses*
1.   Are your hypotheses *clear, concise,* and *specific*?
2.   Are your hypotheses *adequately tied* to previous research?
3.   What *variables* or *concepts* are in your hypotheses?
     Do they have empirical referents? Are they specifically defined?

D.   *Study Area/Spatial Framework*
1.   Have you *clearly defined* your spatial framework?
2.   Where will your study take place? What are the geographic boundaries or the geographic locations of your study? Why?

E.1.   *Methodology/Methods of Study*
1.   Have you *clearly explained* your methodology?
2.   Is it *rational, sound, orderly,* and designed to deal with potential problems, clearly geared toward solving the research problem?

E.2.   *Data Generation*

---

1. We are grateful to Professor N. F. Henry, SUNY-Binghamton, for providing some of the material for this Table.

1. What data are needed to test your hypotheses? What *variables* are in your hypotheses? How will each be *measured?* Make a list.

   Variable   How measured?

2. Where will these data come from? Make a list.

   Variable   Source

3. Are *primary data* required? How will you select observations? Is your *sampling/framework* justified?

4. If primary data are required, is an *instrument* available, or must you purchase or design one?

5. If you have created a survey research instrument (questionnaire), does each question have a *clear purpose, function,* and *relationship* to at least one hypothesis?

E.3. *Data Portrayal and Analysis*

1. How will your sample data be portrayed/illustrated? How will the data be analyzed?

2. What maps and *graphics* are necessary? Will they be essential, useful, or filler?

3. What *cartographic technique(s)* will be used? Why?

4. What *statistical model(s)* will be used? Have you specifically stated the model? What are the data requirements of the statistical technique(s)?

5. What *hardware* and *software* are required for these illustrations/analyses? Are they readily available?

· F. *Timing/Scheduling*

1. Is your proposed time *framework adequate* for the completion of this research?

2. *When* will your study take place? Why? Are there any *confounding situations* associated with this time?

The literature review provides knowledge of the research project and, therefore, the theoretical framework of the proposed study (see Table 3.2, Section B). The student must answer questions about the thoroughness of the review (coverage of the topic), how well the research notes summarize that literature, and what the most important variables and findings of the previous research are.

Table 3.2, Section C reports questions related to hypotheses. They must be conceptually clear and specifically stated in a concise manner. Other important questions include how well linked they are to previous research (theory) and whether or not each concept and variable has empirical measure and meaning and is specifically defined. To be certain that these issues are appropriately addressed, the student should create a list of well-defined concepts included in each hypothesis and its elaboration. Empirical measurement and meaning ensures that moral judgements or other attitudes that cannot be verified by the research are not presented in the hypotheses.

Questions related to the study area also appear in Table 3.2, Section D. We noted in Chapter 2 the importance of determining precisely the areal extent of the research area or matrix because the size of the study area has a direct bearing on "what is studied" and at "what level of detail." The student researcher, therefore, must address the questions of where the study is to take place, and why, as well as be certain that boundaries are clearly defined and justified.

We noted in this chapter that the *project design* describes *how* the research is to be executed. The researcher, then, after successfully stating the research problem (question to be answered) and stating clear, concise hypotheses, identifies the *exact methods* to be utilized in testing the hypotheses and, therefore, in answering the questions posed by the researcher. This is crucial to the orderly execution of the project. It is also important to other researchers who may wish to use your findings, either as a basis for replicating your research or as support for some theoretical proposition related to a new study. In such cases, the researcher must know the specific hypotheses that you tested and the precise methods and tools used to collect and analyze your data and to lead you to specific conclusions. This is a standard requirement of all the sciences and explains why they adhere to the scientific method.

Returning to Table 3.2, Section E.1 poses questions about the overall methodology, or set of procedures, to be used and about the generation, portrayal, and analysis of the data that will be used to test the hypotheses. The expectation is that the researcher provides a rational, sound, and orderly methodology that leads clearly to the solution of the research question.

The research hypotheses dictate the data requirements. Therefore, it is useful to make a list of variables identified in the hypotheses and to state how each variable will be measured (E.2). The same is true regarding the data sources (where you will get the data). If appropriate data are unavailable, then it will be necessary to generate them. Since it is impossible to observe every occurrence of each variable in the study area, it will be necessary to address the question of sampling, or how to choose which observations (phenomena or people) to include in a sample. Also, if primary data are required in the research, questions of the availability and cost related to measuring instruments must be addressed. This could be equipment such as a soil auger or a questionnaire to be administered. If survey research is required, then questions related to questionnaire design and administration must be addressed, including the purpose of each question posed and its specific relationship to at least one hypothesis.

In Table 3.2, Section E.3 raises questions about the portrayal and analysis of the sample data. How will the data be illustrated (graphs, charts, maps)? What statistical tools will be used to analyze the data? The data measurement is linked to the statistical technique selected for the analysis of the sample data. Some form of quantification will be required. The statistical technique may require that the sample data appear in a particular form. For example, if the proposed analysis requires that a mean (average) distance be calculated, say for how far a farmer travels to purchase seed, then a particular type of data is required. The researcher cannot collect categorical data (e.g., 0–5 miles, 6–10 miles, 11–20 miles, more than 20 miles), in this case, but must record the actual numerical value that represents trip distance (so that an average can be calculated).

Finally, we noted earlier in this chapter that timing and scheduling are important to any research design. In Table 3.2, Section F itemizes two questions relevant to this issue. The basic question is whether or not the researcher allots adequate time and resources to the proposed research. This is important to the student researcher who must meet an academic deadline (or suffer the consequences). In the "real" world the issue involves dollars and cents. If the researcher errs estimating the time and resources involved in a research project, the agency stands to lose money, or to face litigation. The personal consequences for the researcher are obvious.

There is another "timing/scheduling" issue. This question involves "when" the proposed research is to take place. A simple example of "timing" involves "seasonality," which can have serious consequences in both environmental and human geographic research. Two examples illustrate the importance of "timing/scheduling."

A researcher proposed to test the hypothesis that run-off in Area X will increase by 15 percent with the construction of a regional mall and a surrounding new residential village. Two test periods for data generation were proposed: June and November. The researcher was aware that rainfall varied for the two months and explained a procedure to account for it. All other things were viewed as "constants" in the environment of the study area. The study area was located in upstate New York. His analysis was presented to a client who criticized the study for ignoring another important variable related to seasonality. Can you guess what it was? Vegetative cover was very different during the two sample months and was unaccounted for. The client refused to pay!

Another researcher proposed to explain the variation in the retail sales of "paper goods," including stationery, greeting cards, and related products. A busy schedule required that sampling of shoppers occur only in two periods: May/June and November/December. These periods were chosen by the researcher as being "like any other two-month periods of the year." Did this researcher adequately consider "seasonality" in the research design? No. You probably quickly recognized that the November/December period is not a "typical" period due to the purchase of paper products associated with the holidays (Thanksgiving, Hanukkah, and Christmas). However, you should note that the May/June period also contains events and holidays that influence paper good sales—graduations, Mothers' and Fathers' Days—and, therefore, does not reflect a representative period. There are ways to adjust for such influences but these must be part of the research design.

The issues in Table 3.2 provide a good basis for questioning all aspects of a research study. They are recommended as a "checklist" for the student researcher. In the remaining chapters we focus on data, their collection, portrayal, and analysis, and on the preparation of reports and proposals.

Acquisition of Relevant Data

Acquisition of geographic data involves several techniques that have been well established for many years. All major university departments of geography offer one or more courses focusing on these techniques and the development of associated skills. In fact, these skills are considered essential to the professional training of a geographer. They are often referred to as "techniques" or "methods" courses, for example, Field Techniques, Cartographic Techniques, and Quantitative Methods.

Whether these skills are referred to as methods, techniques, or tools is not particularly important. They are relative terms, and in a hierarchy of importance, the order would be: (1) method, (2) technique, and (3) tool. In this discussion, the term "method" will be used to describe the general research scheme or framework which determines the kind of result sought—a law, a norm, or a history. The type of research problem determines the "techniques" or skills that must be applied to produce the solution. A technique refers to a design of procedure that best performs the job at hand. For example, a geographer might employ "cartographic techniques" to illustrate the spatial distribution of phenomena. A "tool" is an instrument or device used for skillful technique operation. The geographer uses a wide variety of tools, depending upon the specific nature of the task at hand. Tools include aerial photos, weather instruments, maps, computers, drafting instruments, alidades, and compasses. In addition to such physical tools, mental or conceptual tools, such as a random sample or a mathematical model, are also used.

Like other disciplines, certain basic skills are used in geography, such as library research, logical deduction and induction, and written expression. However, these skills are so much a part of, and so essential to, higher learning that they are taught by departments established for no other purpose.

Special skills required in geographic research are usually taught as geography courses and require considerable time and direction for their mastery. This is true of field techniques, cartographic techniques, air photo interpretation, and quantification. Research work in geography implies that a working knowledge of all these skills has previously been attained, as well as a mastery of one or more techniques pertinent to the specific research task at hand. Frequently, the beginning research student faces some confusion in applying these skills. The courses are generally taught as isolated units. They are taught as skills, which they are, and it is often not clear to the student about when and where a specific technique should be used within the overall structure of the research process. The purpose of this chapter is to establish the order of the research process and the techniques most effective to each stage.

Scientific research is the gathering and processing of data on which scientific truths are based. It is a logical and systematic sequence of related steps. The order of these steps, or stages, is: (1) the collection of data, (2) evaluation of the data, (3) analysis of the data, and (4) prediction based on the analysis. Each stage of the research must be pursued in the proper order, because it is virtually impossible to complete any stage without the knowledge obtained in prior stages.

The ultimate aim of scientific research is to build theory as a basis of understanding and prediction. It might be noted that some research, especially that of an exploratory nature, does not predict but, rather, obtains basic new information that was previously unknown. In situations where data are lacking or difficult to obtain, the gathering and organization of the material may be a major task. Usually, elementary research at the undergraduate level entails only a descriptive report. Most graduate research work, however, and especially a thesis, requires analysis and some form of model development or prediction. Geographic techniques provide the necessary skills for the competent completion of each of these four stages.

Geographic techniques are employed in all stages of the research process. Each specific technique is employed most intensively in one or two stages; and it is well to recognize the contribution of a given technique when planning the overall research design. Table 4.1 indicates the technique for each stage of the research, although they are not necessarily limited to the particular stage shown.

TABLE 4.1

Techniques Used at Various Stages of Geographic Research

Stages	Techniques
I. Data Collecting (Gathering pertinent information.)	A. Library: search and survey for existing relevant published materials. B. Cartographic and Remotely Sensed Data: analysis of existing maps, air photos, and imageries. C. Field: acquisition of new data; mapping and interviewing procedures.
II. Reevaluation of Data (What is the precise nature of the information gathered?)	A. Appraisal of Published Materials: type of information, quantity, source date, quality, and relevancy. B. Evaluating Cartographic Materials: total areal coverage; type of phenomena illustrated: scale, date, accuracy, quality, and relevancy. C. Evaluating Field Data: type of data; how obtained; intensity, scale, accuracy, and quality.
III. Data Analysis (What does the information mean?)	A. Logical Procedures: deduction, induction. B. Statistical Compilation: arrangement of data into tables, diagrams, and models. C. Cartographic Compilation: arrangement of data spatially; construction of maps and space-distance models. D. Correlations and Relationships: determination of causal and non-causal associations, correlations.
IV. Development of Theory or Prediction (What are the results? Can they be projected?)	A. Construction of Models, Laws, and Norms: development and validity of basic models; mathematical, cartographic, logical documentation. B. Projections: predictive accuracy to specific areas, regions, world; temporal restrictions.

It is unlikely that all techniques shown in Table 4.1 will apply to a specific research problem. But science attempts to find truth with "no holds barred," and the researcher may employ whatever skill found useful in the search for a true answer to the problem. Data collecting is essentially making an inventory of pertinent facts. Some of this factual information may be in the form of printed articles, books, maps, and published tables, charts, and statistical data. If so, it is obtained by reading, taking notes, comparing maps, and studying image reproductions. If it is not in published form, the information may be in the "field," thus requiring the utilization of one or more field techniques. This may involve mapping selected phenomena of the visible landscape, either singly or in combination; taking photographs; making notes; and either directly or indirectly interviewing people who have the desired information. The researcher, after collecting the essential data, must appraise and organize them in some logical order so that they may be analyzed in relation to the hypotheses at a later date.

Library Resources

Since the very term research denotes a process of checking and re-checking in order to become certain, it is evident that the library occupies an important position in the process. Probably the advanced geography researcher relies less completely on library sources than does the beginner. Presumably, the former has a better background in the literature of the chosen topic, emphasizes individual reasoning to a greater extent in work, and generates more of the desired data. Even so, both the advanced and the beginning student find the library essential for reviewing the literature and obtaining source material for their studies.

In many cases, data obtained from library sources may be sorted into three broad groups: (1) essential information to the research problem, such as census data where the research is designed for its use, (2) useful data that provide general background information and set the stage for the research problem, and (3) interesting information that relates somewhat to the problem but is not essential nor highly relevant.

There is no standardized first step for beginning the actual search for material written on a geographic topic. However, the library is the depository for recorded information and is a logical starting point. The elementary and background material already known about the topic may be concisely stated in the encyclopedia, an atlas, or a statistical almanac; and the researcher should probably begin by reading these sources. In addition to acquiring background information, a start is made in developing the reference bibliography.

The next step in acquiring information in the library is to survey the card catalog. Here all books are indexed by author, title, and subject. After noting any potentially helpful items in the subject file, the researcher should check authors and titles for any items which may be used in the bibliographic file.

The third step is to search the standard and special references in the library that are available to researchers. In the first category, *Reader's Guide to Periodical Literature, Guide to Reference Books, Basic Reference Sources, Cumulative Book Index,* and the Library of Congress' *National Union Catalog* are suggested. For geography researchers, the indexes to each of the major geographical periodicals should be searched carefully. Documents focusing on comprehensive bibliographic references are available; these include existing lists of reference materials (see Appendix A). For articles in foreign languages, one must refer to the specialized references of the selected area.

The student researcher should make bibliography reference cards of all pertinent published material. In making these note cards, the student should list complete information. It is from these cards that the final material for the printed bibliography is obtained. When using the scientific notation, the cards should appear like the examples below. The space following the bibliographical data is for annotation comments concerning the material in the publication. It is recommended that the cost of the publication be included when available.

Book Example:

Geographer, Geraldine. 1990. *Geography of place:* Dubuque: Wm. C. Brown Company Publishers. (Annotation)

Periodical Example:

Space, Thomas. October, 1990. Political voting patterns: a model for political geography. *The Spatial Enquirer,* pp. 47–55. (Annotation)

The bibliography will expand as reading for substantive material proceeds during the research period. All scholarly articles normally refer to other works that shed additional light on the subject. In fact, such references are so numerous that the researcher will be thankful that there is a clear and well-defined problem. Otherwise, the amount of material referred to is so large that it would be impossible to determine which leads are worth pursuing and which ones should be ignored.

Note Taking

In addition to the bibliography reference card, the researcher must record information on note cards from which he or she eventually writes the report. Proper note taking is absolutely essential for the successful assembling of data for the research project. It begins in earnest immediately after the working outline is completed, and it continues until final organization for writing the research report. As the researcher searches for the specific ideas, facts, and statements that bear on the problem, notes are made of each significant item found. In general, one seeks to record all useful items and avoids taking notes that do not contribute directly to the specific study.

There is no one system of note taking, but every researcher should have some system that is personally meaningful. The card itself may vary in size from three-by-five inches to five-by-eight inches, depending on the desire and the writing style of the user. A basic text should be consulted if the researcher does not have a format in research for a review of methods used. In any event, four items must be known concerning any substance note: the author, title, date, and page of the reference. This does not imply that all of these facts must be shown on the card. For instance, "Doe, *Research Methods,* 17," will lead the researcher to the proper bibliography card if only one work by Doe was used as a reference. Also, "Doe, 1970, 17," is sufficient. Some researchers number their bibliography reference cards as they use them and then record that number on the substance note card. If the book on research methods by Dr. Doe was the third one read, a number "3" would be placed on the bibliography reference card. Then the item recorded on the note card could be labeled "3, 17." In writing the report, a reference to "3, 17" would direct the writer to the work of Dr. Doe, and it could then be properly footnoted. It should be kept in mind, however, that the bibliography reference cards must have the complete name of the author, the title, publisher, publication location, and date of publication.

The material on the note card should be carefully selected. The researcher should never put more than one item on a card. The card should be made either at the particular time an item worthy of note is read or at the completion of the reading of the entire work. This last method saves some worthless note taking and interruption in reading. If it is used, some method of indicating the location within the article of the important item to be recorded must be devised.

Notes may be categorized into five groups: (1) *the précis,* a useful recording of a thought in the researcher's own words, (2) *summary notes,* condensations of pertinent information in the article, (3) *paraphrase notes,* similar to a **précis** but more detailed, (4) *critical notes,* which evaluate and make critical comments on the material; and (5) *quotation notes,* to be used when a statement is so appropriate that it is wanted for inclusion, and the researcher cannot devise a better way to state the thoughts in his or her own words. Needless to say, great care is required in taking quotations accurately and reproducing them in the paper in proper context.

Reference footnote form varies, and the researcher should follow the format designated by the local department or institution. A common form is to place a number at the completion of the material to be noted (raised one-half space) and to show the reference at the bottom of the page in the same way that footnotes in this book have been treated. A shorter method, one which is more efficient and is becoming widely accepted, is the inclusion of the reference immediately following the material to be noted. The following is an example: (Geographer, 1970, 21). The important consideration is that the reader of the research report be given sufficient information regarding the reference in order to check it readily in its original form. The information tells the researcher who said it and when. He may locate the source from the bibliography and investigate the quality and reliability of the reference.

Appraisal of Library Source Material

Judging or evaluating source materials is part of the training of the research scholar. The value of the completed research work depends in large measure upon the quality of the sources used. Evaluation of published materials normally includes five components. First, what is the source of the original data? Was it obtained by interviewing, mapping, aerial photo interpretation, or a combination of two or more of these techniques? Second, what is the scale of the data? The scale determines the degree of generalization and is this generalization suitable to the problem? Third, what is the date of the material? In some research problems concerned with dynamic areas the date is critical. Fourth, how was the data classified? What phenomena were selected, how were they grouped or generalized, and for what purpose? Finally, who was the responsible individual, group, or agency that compiled the information and for what reason or reasons? Such an appraisal will determine to a large degree the validity of library materials.

Source materials may be classified as primary, secondary, and tertiary. Often, a particular work does not automatically fall into one category or the other but varies in relation to the nature of the study. For example, a textbook is usually considered a tertiary source because it is compiled mostly from secondary sources; but it may be a primary source in a study in which the major objective is to determine how textbooks treat the subject of model building.

Primary sources are usually considered the best sources in research. Such source material is near its original form and is relatively free of editing, alteration, or modification. As such, it tends to be divorced from external influence, judgment, and bias of others which might lead to unsound interpretation by the researcher. Primary sources, then, are original descriptions or analyses of a process or event. In this category, one commonly finds data concerning experiments, interviews, questionnaires, field studies, letters, diaries, autobiographies, creative works, and statistical reports. No interpretation by the collector of these sources is attempted.

Secondary sources are usually factual accounts written about a subject. The information commonly represents selected data taken from primary sources that have been organized and interpreted by the writer. A primary source may contain some secondary source material. For instance, a field study may include material interpreted from previous field studies and analyzed by the field reporter. This analysis then becomes a secondary source, as it includes a judgment factor. Only new information directly attributed to the field researcher's own efforts is primary in nature.

Tertiary works are compiled from secondary source material. Most textbooks are in this category. Well-conceived tertiary works become widely accepted and serve as standard reference documents.

A source may be primary, the interpretation absolutely correct—and still be of little value. Its use in such cases would be an indication of poor scholarship on the part of the researcher who is judged on the critical evaluation of each source cited. Obviously, any position could be supported by citing someone. For example, a recent political analysis in a national news magazine directly quoted twelve sources. This list of implied political "authorities" included an aluminum plant worker, a printer, a boxing manager, two pollsters, an unnamed "moderate Republican senator," and a "White House advisor." Two glaring research errors are obvious: lack of precise identification of the "authorities" and lack of any indication of their competence.

The research student must also exercise discrimination in the quality of the sources; this is based on the ability to judge each work. Some of the criteria to be applied in such judgement is external. In this group, such things as the reputation of the author and the publisher, and the comments of critical reviews and annotated reference works are considered.

Internal criteria require more discriminating skill. The researcher will wish to arrive at some evaluation of the quality of the work before investing too much time in it. The usefulness of the material may be evaluated by reading the introduction, the purpose, or the preface to determine if the intent of the author is to treat the subject in a way that appears satisfactory. The date is evidence of the chronological pertinence of the study, and any comments about the author indicate something of the author's reliability. A sample reading of a few pages will suggest the level of reasoning in the work, as well as the clarity of the writing. At the same time, footnotes can be checked for accuracy and adequacy. The bibliography of sources used, whether lengthy or short, is an indication of the quality of the work.

Plagiarism

The "coin" of the scholar is the researcher's scholarly creations. Plagiarism is the use of these creations without giving credit to and/or getting proper permission from their creator. Such creations include not only the exact words of the originator, but also his ideas, phraseology, and original organization of materials. Such originals are protected by law, and in serious cases of usurpation, legal liability can be established.

Most plagiarism is unintentional and results from errors in note taking or reporting. Even so, such errors are not to be treated lightly because they indicate lax scholarship. It is not difficult to avoid this pitfall if extreme care is taken to give proper credit. When any author's copyrighted material is used, there is a possibility that this use will compete with the original, and so permission should be obtained. The best single criterion to keep in mind is: "Does use of the material impair the value of the original?" Such interpretation may be difficult, and if too strictly followed, could preclude the use of almost all sources. The writer should use reason, and when in doubt, risk erring on the conservative side. The length of the material used is not a criterion for determining plagiarism. Each situation differs, and there is no rule of thumb to serve as a guide in this matter. Most types of creative work may be protected by a copyright. A potential author should be familiar with the regulations.[1]

Field Techniques

In many research problems, key information or essential data may not exist in published form. This is particularly true if the research area is small and detailed data are required. It then becomes necessary to obtain information by employing one or more field techniques. Geographic field techniques essentially consist of recording direct observations within a well-conceived and scientific framework. They provide a structure that enables the researcher to make order out of chaos (see Figure 4.1).

---

1. United States Copyright Office, *Copyright Law of the United States of America* (Washington, D.C.: U.S. Government Printing Office, 1967).

FIGURE 4.1

Conceptual structure of geographic field methods.

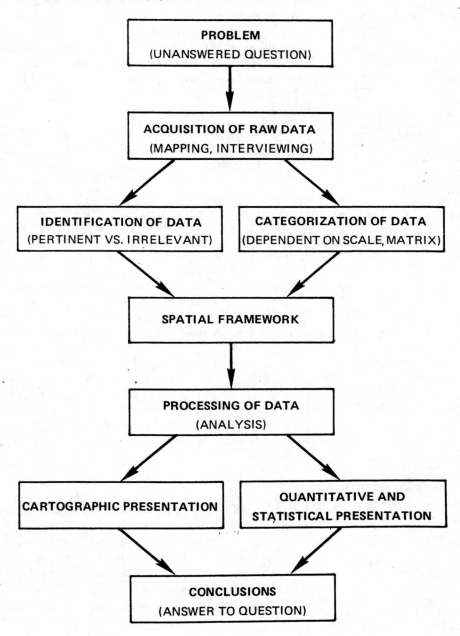

As is true in other sciences, the techniques and research equipment used by geographers to collect raw data have evolved over the years. Direct observation as a major source of data is deeply ingrained in the history of geographic thought. Indeed, it was less than 100 years ago that a "geographer" was one who journeyed to exotic regions and returned to recount travel experiences, including general observations made. In the event that a descriptive report was presented before a prestigious group, such as the Royal Geographical Society of London, the geographer had reached the apex of his career. Today, direct observations can be accurately recorded and scientifically structured. Many geographical studies lack sufficient and basic data without the information provided by modern field work. Field work is the collection of raw data by primarily using: (1) mapping techniques and (2) interviewing techniques. Most field work employs both of these techniques because one implements or supplements the other.[2] This chapter focuses on mapping techniques and Chapter 5 is concerned with survey research procedures.

Mapping Techniques

Mapping raw data is concerned with the spatial recording of visible features, or phenomena, of the landscape that may be categorized. Field mapping consists of several distinct steps which must be completed in order. These steps are: (1) determination of the *scale* of the mapping task, (2) *classification* of the data to be mapped, (3) selection of the *base map* upon which the data will be recorded, and (4) the *actual field mapping*.

The selection of the scale determines the size of the minimum mapping unit. If the research matrix is small and detailed data is required, the minimum mapping unit may be only a few square feet in size. On the other hand, if the research problem is concerned with a large matrix and data are less detailed in nature, the minimum mapping unit may be several acres in size. The scale of mapping determines to a large degree the way in which data are classified and the precision of the actual field mapping.

A map is a human-made representation, drawn to scale, of a segment of the earth's surface. It is always smaller than the part of the surface which it represents, and so it is not practical to show each and every aspect of the total landscape exactly as it occurs. Certain features, or phenomena, must be selected for recording, and those selected must

---

2. For detailed information concerning field methods see: Lounsbury and Aldrich: *Introduction to Geographic Field Methods and Techniques,* 2nd Ed. (Columbus: Charles E. Merrill Publishing Co., 1986).

be generalized, or classified, in some meaningful manner. The process of classifying, or categorizing, data requires serious thought. The type of classification bears directly on the outcome of the research problem. The need to classify and categorize is not unique to the field of geography. In all sciences, physical or social, the development of taxonomies, or classifications, is prerequisite to standardized research. The orderly accumulation of information leads to the formulation of conceptual frameworks and basic laws. The classifications of elements, plants, animals, and rocks, as well as of human activities and events, are a few examples of a multitude. The existence of many recognized subfields within the discipline of geography requires the general geographer to have a working knowledge of many classifications, and the research geographer must have the ability to conceive sound and scientific classification systems as the need arises.

Mapping implies recording facts, drawing boundaries, and making notes on *something*. This something must have reference points or control lines accurately placed on it so that the mapper knows where he or she is at a given time and can accurately locate features of the landscape on the field or base map. A base map may be looked upon as being a type of notebook upon which facts and observations are recorded. However, unlike a notebook, the observations, or facts, not only can be recorded, but their spatial or areal extent can be accurately defined. If the classification system of the data being collected is sound, the completed map becomes a scientific document capable of showing a wide variety of pertinent information with many uses and applications. A finished map is essentially a table that presents data spatially.

There are many types of maps and map-like devices that are widely used as base maps for field mapping. They include vertical aerial photos, topographic maps, plat or cadastral maps, air charts, and even a sheet of paper on which some control or reference points have been placed. The selection of the most appropriate base map depends upon the type of data being collected and upon the scale or size of the area being mapped.

Aerial photos are widely used today as base "maps" because they are camera images of a part of the earth's surface. All visible features are shown—crops, houses, trees, gullies, roads, and so on. The aerial photo in itself provides a great deal of information, such as the areal extent of cultivated land, landforms, forests, settlement, and transportation patterns. An experienced researcher may compile land use data and construct a land use map from the photo itself, using interpretation techniques verified by spot checks in the field. A photo is not really a map—it is a picture of all aspects of the landscape—no selection process has taken place. Conventional photographs show everything, although filtering devices and recent techniques of remote sensing allow some selection of features to be reproduced.

Photographing and interpretation of aerial photos is essentially an art that has been developed since World War I. As a geographic research skill, it attracted little attention until 1943. At the present time, in addition to conventional photos, geographers utilize infrared, radar, and other forms of remote imagery. These devices are of major concern to the research geographer because they provide a mechanism to obtain and describe data.

Oblique aerial photographs are taken when the optical axis of the camera is inclined from the vertical. Oblique photos, when used singly, distort surface features and are of little value for mapping purposes where true area and shape of features are needed. Vertical photos, on the other hand, are those taken by a camera pointing vertically downward, and they illustrate areas and ground objects as they actually exist. If, however, the camera is tilted slightly from the vertical, or if there is a great deal of topographic relief, some distortion will be present. In all cases, the edges of the photo are distorted to some degree, and only the central portions of a vertical photo truly represent the surface of the ground. However, despite these disadvantages, vertical photos serve as excellent base "maps" because of the detail shown and the large number of control points that are not normally present on other types of maps. Most parts of the world have been photographed one or more times by government agencies or by commercial concerns. Every county in the United States has photographic coverage, and photos are available in various scales (see Figure 4.2 and Appendix A).

A conventional aerial photograph translates visible light, reflected from the surface of the earth, into a black-and-white image. In recent years, sophisticated instruments have been developed to detect electromagnetic energy invisible to the human senses. This information is often processed in order to present an enhanced color or black-and-white image. These instruments, usually installed on aircraft or satellites, can acquire data concerning the earth's surface from considerable distance and provide information about nonvisible objectives, features, and phenomena that otherwise would go unnoticed. Such data collected by remote sensing techniques are of great value to scientists in a variety of disciplines concerned with physical and human-made environments.

In general the nonvisible data acquired is electromagnetic energy radiated or emitted from objects above absolute zero (0°K or −459.69°F) or *reflected energy* (sunlight, laser light, or microwave radar) reflected from objects or features. All objects above absolute zero radiate energy, and the amount radiated and the average frequency increases as the temperature of the target increases. This directly emitted radiation can be translated into an image that can be analyzed to distinguish various water temperatures, geothermal areas, forest fires, etc. The reflection of visible and nonvisible light may be detected by near-visible infrared. With the use of color infrared film, and when analyzed in terms of the

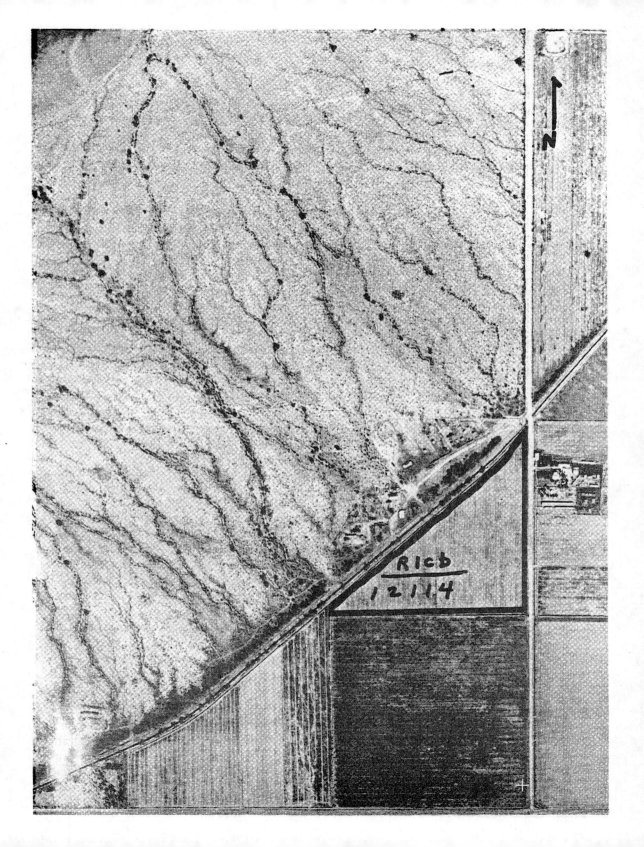

FIGURE 4.2

A portion of vertical aerial photograph (scale: 1 inch = 1,000 feet). Note the presence of field patterns, small topographic, vegetative, and cultural features that are not normally shown on maps. Aerial photographs for all parts of the country may be obtained from the United States Department of Agriculture, Agricultural Stabilization and Conservation Service (ASCS), Washington, D.C. 20250. Each county ASCS Office has an index of the photographs giving the code numbers and date of flight for individual photos in the county. These code numbers must be obtained, as well as an order blank, before photographs can be purchased. Land use mapping may be easily accomplished using vertical aerial photos as base maps. The example below illustrates multifeatured mapping utilizing the fractional code system (details concerning the classification system of land use and physical features used in this example may be found in Appendixes B and C).

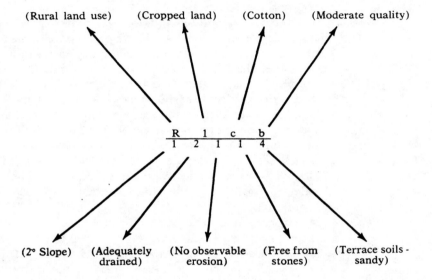

proportion of each wavelength reflected, this provides identification information (spectral signature) to distinguish vegetation types and land use. The reflection of laser light and microwave radar may also be utilized to record relief and topographical features.[3]

Using several cameras and/or sensors, each designed to detect wavelengths of a certain band, a multispectral survey may be taken. These various images of the same area may be combined to produce a single-color photo. Essentially this is the system utilized by the Earth Resources Technology Satellite (LANDSAT), which employs four multispectral scanners and three television cameras from an altitude of over 600 miles in space.

The variety and amounts of data provided by remote sensing techniques are tremendous. At this stage of development, most of the data are for very large areas and detailed information for small areas is lacking. The imagery is valuable in that it provides general information concerning patterns of land use, but is not suitable for reasons of cost and scale for detailed field work or analysis of microareas.

Topographic maps are also widely used as base maps. They are large-scale maps and are available for many parts of the country. They show the exact locations of specific physical and cultural features, which serve as reference points in field mapping. Like all maps, they do not show all of the landscape, because certain features have been selected for illustration, while other features have been omitted (see Figure 4.3).

In collecting urban land use information, plat maps are often used. These maps are concerned only with property boundaries and show the dimensions of properties and lots in feet. They are available at large scales in most urbanized areas of the country. Inasmuch as distances are precisely given on the maps, compilation of the square footage of each type of land use is easy to compute (see Figure 4.4).

In certain situations, a *computer mapping* program, or a *geographical information system,* may provide adequate base maps. Further, the researcher may realize a tremendous savings of time in compiling data at a later stage of the research project. A basic requirement for proper base maps is to have a suitable geocoded base file (GBF) which is the outline of the minimum mapping units within the research area. The GBF, or cartographic base file, must be in a suitable form (computer disk, magnetic tape, or CDROM disk). If a proper GBF is not readily available, its preparation could be highly time consuming. On the other hand, it normally needs to be produced only once for a given study

---

3. For detailed information concerning remote sensing see: Benjamin F. Richason, Jr., Editor, *Introduction to Remote Sensing of the Environment* (Dubuque: Kendall/Hunt Publishing Co., 1978), and T. M. Littlesand and R. W. Kiefer, *Remote Sensing and Image Interpretation* (New York: John Wiley and Sons, 1987).

area and can be used time after time. As the study areas change and new data become available, up-to-date computer generated maps can be produced quickly. Many GBFs were made available from the United States Bureau of Census information (county outlines, metropolitan areas, and other political and demographic units) and in 1990 a product termed "TIGER" combined Census and USGS maps in a "topologically integrated" file for every county in the United States.

There are many types of computer maps and graphics. The researcher may, however, be limited depending upon the type and degree of sophistication of the equipment available at a given institution or agency. The type of equipment has a direct bearing on the cartographic products to be used for analysis or presentation. Commonly used computer graphic and cartographic outputs include standard line printers, dot matrix line printers, pen-and-ink plotters, electrostatic plotters, microfilm plotters, and cathode ray tubes. Each has its advantages and limitations, and each may be suitable for a particular type of research.

There are available computer mapping programs that may be appropriate to certain types of research. Many universities and agencies sell programs they have developed and continue to maintain. There is a wide variety of equipment, programs, and production methods. The researcher should determine precisely what is available at his or her institution and carefully consider whether or not computer mapping is appropriate to his or her research. Computer-generated maps and other graphics are not applicable to all geographic research projects and, in other cases, the cost in time and effort may be too great for the benefits derived. In all cases, however, the researcher needs to assess and evaluate this potentially valuable tool in terms of providing suitable base maps, methods of analysis, modes of presentation, or combination thereof, as it relates directly to a specific research project. Because *computer mapping* and *geographical information systems* have revolutionized a significant portion of geographic research, Chapter 7 provides an introduction to their use.

In the event that no maps of the research area exist at a scale necessary for field-mapping purposes, the researcher, by using relatively inexpensive equipment, may construct his or her own base map. A piece of paper, a compass, an open-sight alidade (sighting device), a plane table with a tripod, a level, and a tape measure are all that are required. The researcher then can construct a series of known points and lines, each in its proper direction and distance from the others, which will serve as a controlling framework within which specific phenomena may be accurately located. Plane-table mapping is not commonly used at the intermediate or macroscale, as base maps for most areas are available. However, in microarea studies, plane-table mapping may be necessary to obtain the precision and accuracy desired.

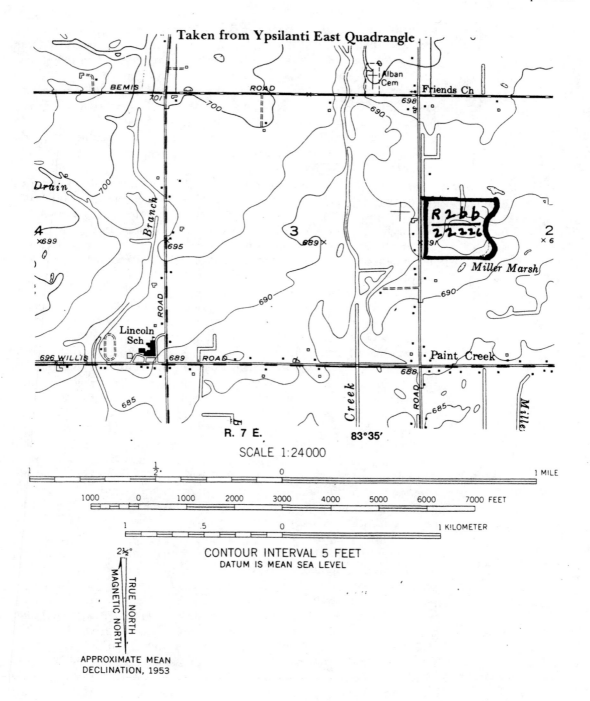

Taken from Ypsilanti East Quadrangle

R. 7 E.   83°35'

SCALE 1:24000

CONTOUR INTERVAL 5 FEET
DATUM IS MEAN SEA LEVEL

2½°
TRUE NORTH
MAGNETIC NORTH

APPROXIMATE MEAN
DECLINATION, 1953

FIGURE 4.3

A portion of a topographical quadrangle, 7 ½′ × 7 ½′ Series, Scale: 1 inch = 2,000 feet (Courtesy of the U.S. Department of Interior, Geological Survey). Note the absence of small ground features found on a vertical aerial photo. However, topographical maps show political boundaries, systems of coordinates, and they identify major cultural and topographic features which may be used as control points in field mapping. The area outlined on the map illustrates how topographic maps may be used as base maps for field work. The fractional code system of mapping and the classification systems are explained in appendixes B and C.

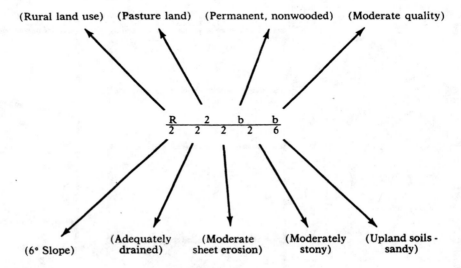

(Rural land use)   (Pasture land)   (Permanent, nonwooded)   (Moderate quality)

$$\frac{R}{2} \quad \frac{2}{2} \quad \frac{b}{2} \quad \frac{b}{2} \quad \frac{b}{6}$$

(6° Slope)   (Adequately drained)   (Moderate sheet erosion)   (Moderately stony)   (Upland soils - sandy)

FIGURE 4.4

A portion of a plat map. These maps are often used as base maps for urban land use mapping. Property boundaries and exact distances are shown, and the precise square footage of each type of land use can be easily computed. Plat maps may be obtained from city or county government offices. This example illustrates a two-family residence in moderation condition, constructed since 1960 (see Urban Land Use Classification, Appendix C).

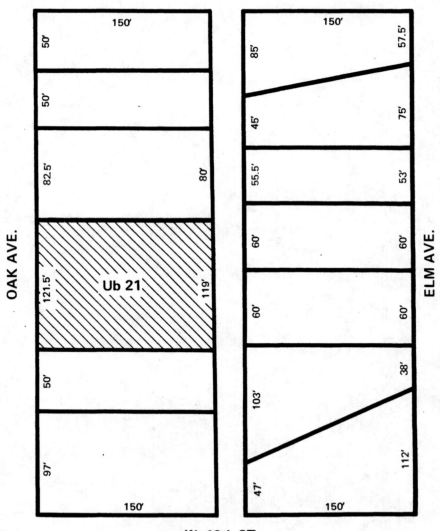

W. 9th ST.

OAK AVE.

ELM AVE.

W. 10th ST.

In constructing a plane-table map, the basic principle is to determine the direction and distance from one selected point to another. The plane table must be leveled, oriented to a compass direction, and a line drawn toward magnetic north. The first mapping location (Station A) is marked on the paper, care being taken to place this location in the proper section or area of the anticipated map when completed. Station B may be a natural feature or a placed rod or pole at any convenient distance. Carefully, a sighting is made and a line drawn along the sighting device from Station A in the direction of Station B. Measuring the distance between the stations can then be translated to the mapping scale and Station B can then be determined on the map. Two known points are now on the base map as well as a line of traverse between them. The mapper then moves to Station B, levels and orients the plane table, selects a location for Station C, sights and records the direction from Station B, and measures and records the distance between Stations B and C. When two or more stations have been plotted, additional points may be recorded by triangulation.[4] That is, sighting on a specific point from two or more known stations and the intersection of lines will determine the location of the points on the plane-table map (see Figure 4.5).

Exceptionally accurate maps may be constructed utilizing the above principles and a telescopic alidade. This instrument is used for highly precise surveying work such as the microanalysis of vegetation and soil types, field and crop boundaries, building locations, lot lines, etc. The telescopic alidade differs from the simple open-sight alidade in that it has a telescope containing a cross-hair reticule, one or more levels, a compass needle, and graduated arcs for measuring the inclination of the telescope. Plane-table maps utilizing the telescopic alidade are more precise than those constructed with the open-sight alidade. The optically powered sighting device provides a greater degree of accuracy in direction determination and distances can be measured by determining the parallel stadia cross hairs contained in the telescope. A tape measure or chain is not necessary. Further, differences in elevation from one point to another can be calculated accurately and topographic features can be mapped (see Figure 4.6).

After an appropriate base map has been acquired and the type and classification of data to be obtained have been determined, the researcher is ready to begin the actual field mapping. Obviously, the mapper must be able to locate ground objects precisely on the base map; and if a sufficient number of features may be identified on the base map, he may then draw the boundaries around whatever phenomena concern him. The student researcher is strongly advised that, when

---

4. For detailed information concerning plane-table mapping see: David Greenhood, *Mapping* (Chicago: The University of Chicago Press, 1964), pp. 203–239.

FIGURE 4.5

A sample plane-table base map (scale: 1 inch = 100 feet). A series of control points has been established, as well as a line of traverse. From this framework, other control points may be accurately located and the areal extent of phenomena mapped. From each station of observation, the exact direction and distance to a subsequent station must be determined precisely if a base map is to be accurate.

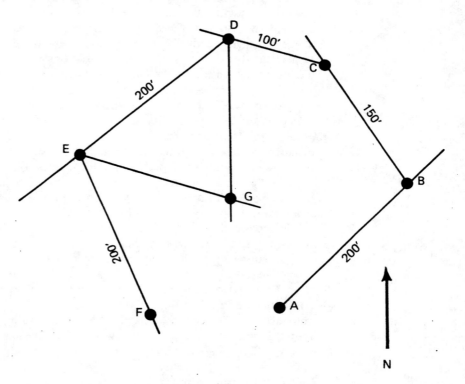

gathering raw data, to collect more than the researcher thinks is pertinent at the time, and that he utilize the most detailed system of data classification that is possible at a given scale. It is not an uncommon practice during the actual field mapping process to identify additional pertinent data that were not recognized previously. Often, the value of this additional information is not fully realized until the data are analyzed at a later stage of the research problem. Multifeature mapping is also strongly recommended, because the resulting information may throw light on unexpected areal associations. When the researcher is collecting raw data, it is not always possible to anticipate in advance the value of the data and the new concepts or hypotheses that might evolve.

FIGURE 4.6

Determining distance and differences in elevation using a telescopic alidade. Distance between two points may be determined by calculating the stadia interval (the number of stadia or graduations on the stadia rod between the upper and lower parallel stadia wires or cross hairs). The stadia wires are fixed and spaced so that the distance is 100 times that of the stadia interval. For example, the distance from Station A to rod location 1 is 150 feet (1 1/2 feet between the lower and upper stadia wires). The distance between Station A and rod location 2 is 200 feet. Differences in elevation between one point and another are simply a matter of level sight readings. For example, if rod location 1 is known to be 850 feet, rod location 2 is 855.5 feet and the height of the instrument is 858 feet and 864.5 feet at Stations A and B respectively. In areas of higher relief, where the rod may be above or below the level of sighting, several methods, utilizing the graduated arcs of the alidade, may be used to calculate elevation.[5]

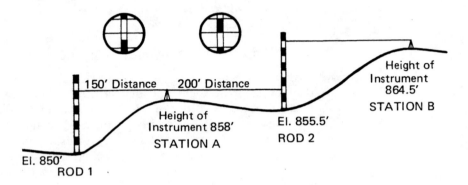

The student who engages in field mapping is well advised to follow a carefully detailed plan of action. Field work is a time-consuming research technique. However, a well-conceived field plan will prevent much wasted effort and time. Fundamental to good planning is: (1) a clear statement of objective(s), (2) a careful pre-field study of existing documentary materials, (3) the proper preparation of base maps and related materials, (4) a reconnaissance of the research area, (5) the field mapping, taking extreme care to record data accurately, and (6) a post-field study to ink in penciled lines, check for gaps, assemble field maps, organize field notes and on-site photographs, and evaluate any unexpected information that has been revealed.

There are many ways that data may be recorded on base maps. For example, *single feature* mapping is concerned with one phenomenon only, such as house types, or land use, or road patterns, as well as a host of others. *Multifeature* mapping refers to mapping several phenomena at one time, such as soil types, along with land use, slope,

5. For detailed information concerning mapping with a telescopic alidade see: Low, Julian W., *Plane Table Mapping* (New York: Harper and Row, Publishers, 1952) or Kissam, Philip, *Surveying for Civil Engineers* (New York: McGraw-Hill Book Company, 1956), pp. 167–198.

settlement patterns, and the like. For this type of mapping, some sort of a fractional code or similar system must be devised (see Figures 4.2, 4.3, and 4.4). *Dimensional* or profile mapping is a widely used technique in urban fieldwork when data from upper floors of high-rise buildings are required. Usually each floor is mapped as one story on a plat map or graph paper in vertical dimension. *Sequential comparisons* are used to measure change. For example, an existing map compiled from the field may be compared to older photos or topographic maps to measure the change that has taken place over a period of time in land use, settlement patterns, landforms, or any dynamic or changing phenomenon. These techniques are only a few of many that are commonly utilized, and new or modified techniques can be devised to suit the specific research problem.

Selection of Base Maps and Mapping Techniques

It is obvious from the previous discussion that there is no one base map or mapping technique that is best for all types of research problems. The nature of the problem and scale of the necessary data determine the mapping procedures to be used. If the problem is concerned with agricultural and rural land use, for example, a vertical aerial photo and multifeature mapping would be the best tools and mechanisms to be employed. Conversely, if the problem deals with central business district delimitations, a plat map, dimension and multifeature mapping would be most desirable. An important and integral part of the overall research design is the selection of the most appropriate base maps and mapping techniques.

CHAPTER FIVE

Data Generation by Survey Research

.

In many research problems, the necessary raw data may not be visible and, therefore, cannot be mapped in the field. For example, in an agricultural study, the researcher must gather information concerning the marketing of products, rotation of crops, types and amounts of fertilizers used, labor requirements, and other related variables to gain a complete understanding of the agricultural geography of the area. For these data the researcher employs survey research techniques seeking *reliable, precise,* and *accurate* data.

Reliability is the degree to which other researchers, applying the same measurement devices to identical phenomena, obtain similar results. *Precision* refers to the fineness of the measurement scale. *Accuracy* reflects the degree to which what is purportedly measured is actually measured.

Achieving accurate data from survey research is a major challenge for a number of reasons. The researcher must try to control several potential problems. First, the survey respondents may not provide accurate. information. Second, there may be a flawed sample and/or a flawed survey instrument. Finally, human error, such as a recording error, can occur. Examples of these errors include: (1) the interviewee does not understand a question and provides erroneous information (questionnaire design problem), (2) the survey did not reach a representative sample population (sampling design problem), (3) the person interviewed intentionally provides misleading information, (4) the language of a survey leads the respondent to give (or withhold) a particular answer (survey design problem), and (5) the interviewer may not comprehend the respondent's answer and records it inaccurately (human error).

The accuracy of the information acquired by the survey research process is dependent upon:

1. The sampling procedure, if everyone in the study area is not to be surveyed;
2. The structure and design of the questionnaire;
3. The way in which the questionnaire is administered; and
4. The way that the responses are recorded, compiled, and tabulated.

In the following sections, we elaborate on these determinants of accuracy.

Sampling Decisions

A geographic researcher often must base research findings on data that are "representative" of some total population (universe). For example, the physical geographer or geomorphologist seeking explanations of variable stream discharges cannot possibly examine all streams for the velocity, depth, and other important parameters that contribute to the problem. The researcher must decide which streams to include in the study, determine which variables will be examined (velocity, depth, etc.), and then determine how many measurements must be taken in each stream in the sample.

Human geographers are faced with the similar dilemmas. While there are many issues to raise about a research question, here we limit our questions to "which" rural households and "how many" households. For example, a discount department store chain wished to answer the following research question, "What proportion of rural consumers will travel at least twenty miles to purchase hardware products?" Clearly, the chain could not ask every rural consumer this question. Like many research questions, this one examined a large and dispersed population. Therefore, the researcher must decide which rural consumers will be asked the question, with the goal of generalizing the results to all of the rural population.

In terms of "which" households are to be asked the question, the researcher must decide the research matrix. Does the chain wish to generalize in this case to all of North America, the United States, or some particular region? In this case, the retail chain wishes to narrow its matrix to the rural area that is defined by geographic uniformity and rural character, the "Northern Tier" of Pennsylvania. This region is defined by its location in the Appalachians and is sandwiched between four major metropolitan areas. This area is of a formal rural development district characterized by very small towns (no town exceeds 10,000 population) and a very low population density (43 people per square mile). This large rural region becomes the "universe" from which the researcher must select a certain number of households to be "representative" of all the rural households of this region, which total 60,000.

When collecting survey research data, the typically large size of the research matrix prohibits total coverage of the area (the goal). Unable to gain total coverage, the researcher devises a system of sampling. The main questions are: How many households are asked the question? Which households are asked the question? The first question is quantitative. Once this question is answered, the second question involves selecting the individual households that will constitute the sample population.

Establishing Sample Size

A sample is some number of observations drawn from and, therefore, smaller than the universe, on whom measurements have been (or will be) secured. Since the sample entails a relatively small number of persons (or elements), the researcher wants to be very certain that the selected individuals truly represent the universe (the entire population). The size of the sample and how it is drawn determine how well it represents the universe. The sample size is based on how reliable the final answer must be. In practice, time and money constraints often play a dominant role in choosing the sample size. Experts, such as Sheskin (1985),[1] have noted that the costs related to survey research are both fixed (e.g., questionnaire design and sampling strategy) and variable (e.g., labor, postage, printing) and that "an important relationship exists between costs and sample size" (Sheskin, p. 31). Similarly, a serious time constraint, such as the need to meet competition in the market, can cause a trade-off between the reliability of an answer and the sample size. Thus, in practice, the ideal sample size may be compromised by time and dollar considerations.

Although trade-offs are common in establishing sample size, setting an acceptable level of accuracy for survey research results should be the first step in sample design. Sheskin provides an excellent explanation of the quantitative procedure utilized to determine sample size for a large population:

---

1. I. M. Sheskin, *Survey Research for Geographers* (Washington, DC: Association of American Geographers, 1985).

"It is common for surveys to report both percentage responses and numerical averages. The nature of these results also affect sample size. The most commonly-selected accuracy level is the situation in which one is 95% certain that no estimated percentage is off by more than +/–5%. (referred to as '95 and 5'). 95% is the confidence level; 5% is the confidence interval. The sample size **n** necessary to achieve such a result is computed as follows:

$$n = (\frac{Z\sqrt{PQ}}{C})^2$$

where   **Z** =   1.96, for 95% confidence that a result lies within
                         a given confidence interval

           Z =   2.58, for 99% confidence that a result lies within
                         a given confidence interval

           P =   the percentage about which a confidence
                         interval is computed, expressed as a proportion

           Q =   1 – P

           C =   the desired size of the confidence interval,
                         expressed as a decimal number."[2]

Sheskin and others have noted that in the "worst-case scenario" the maximum **n** occurs when **P** and **Q** each are equal to 0.5. As a result it is common to use 0.25 ($\sqrt{P \times Q}$) in the equation. The sample size required (assuming the other parameters are constant) is determined not by the size of the universe, but by the desired confidence level and acceptable error. Thus, a universe of 250,000 requires the same sample size as a universe of 500,000 to achieve a given level of confidence with an accepted error. We can construct a table of sample sizes required for specific confidence levels and error ranges. Such a list appears in Table 5.1 (after Sheskin see footnote 1).

If our retail chain conducted their survey and found that 52 percent of all rural consumers would travel twenty miles to purchase hardware, the retailer can be 95 percent certain that the estimated percentage (52 percent) does not err (higher or lower) by more than 5 percent. The chain can be comfortable that the percentage of rural consumers willing to travel that distance for hardware purchases is not less than 47 percent and not more than 57 percent. If for some reason the retailer is not satisfied with this *range* of error, then the sample size may be expanded to narrow the range. For example, if the chain demanded the same confidence level but a 2 percent error range, then the sample size must be 2,401 people. Clearly, confidence level and acceptable error range influence sample size and, the cost in time and money to con-

2. I. M. Sheskin, *Survey Research for Geographers* (Washington, DC: Association of American Geographers, Resource Papers, 1985) p. 33.

Table 5.1[3]

Sample Sizes Necessary to Achieve Various Confidence Intervals[3]

Range of Acceptable Error (±%)	Confidence Level	
	99%	95%
1%	16,587	9,604
2%	4,147	2,401
3%	1,843	1,067
4%	1,037	600
5%	663	384
6%	461	267
7%	339	196
8%	259	150
9%	205	119
10%	166	96

[a](Assumes P = Q = .5 and a very large population)

duct a survey. The researcher must choose an appropriate sample size (accuracy, money, and time) and be able to live with the results.

There are other considerations that require larger samples. We direct student researchers who plan to develop sampling procedures to study the bibliographic references provided in Appendix A. We now turn to the second question, whom to sample.

Selecting Sample Elements: Types of Sampling

Drawing a sample involves two important issues. First, the researcher must identify the source of the sampling universe (a telephone book, a list of commercial establishments, a list of university students, a city directory, or a commercially-produced cross-reference directory).[4] The quality of the source will affect the quality of the sample. The source must assure complete coverage of the research area and the target population.

Second, to be taken seriously by other professionals and to allow the use of statistical inference, a scientific sample must be taken. Types of scientific sampling procedures range from random sampling to more complex methods that involve stratification and/or clustering of population elements. The sampling technique selected is based upon research goals and the knowledge of the researcher about the universe to be studied.

3. Taken directly from I. M. Sheskin, *Survey Research for Geographers* (Washington, D.C.: Association of American Geographers, Resource Paper Series, 1985), p. 35.

4. For example, a Hill-Donnelley Corporation *Cross-Reference Directory* for a particular city, which contains names, mailing addresses, and phone numbers.

Random Sampling and Systematic Sampling

In order for statistical inference to be valid, our sample population must truly represent the total population (the universe). A random sample gives every individual in the universe (object, household, person) an equal chance of being selected. Theoretically, the distribution of characteristics of the sample population corresponds to the characteristics of the universe. To randomly select a sample every individual in the universe is assigned a discrete number. The researcher, employing a random numbers table (see Appendix F), selects numbers that correspond to those in the universe until the predetermined sample goal is reached. Only individuals (objects, households) whose numbers are taken from the random numbers table are included in the sample.

The time and cost required to complete a purely random sample are high and for these reasons it may not be used. However, it offers the advantage of a true representation of the population (universe), when the researcher *knows little or nothing about that population*.

A more frequently-used method is systematic random sampling. While not the equivalent of pure random sampling, its ease of use makes it a practical favorite. In this case, the first observation is randomly chosen from the population (universe), then every n^{th} subject is selected until the desired number of observations is researched. For example, a researcher wants to determine the average price paid for a pound of tobacco to farmers in a homogeneous tobacco farming area of 575 farmers. At the 95 percent confidence level with a ±10 percent error, 96 observations are necessary. The first farmer is randomly selected from the list of 575, then every sixth farmer after that is selected. This would save the researcher a great deal of time and, therefore, money.

A simple random sample is appropriate when we have no prior knowledge of the population. This technique, while being representative, does not guarantee that the individuals selected are from all parts of the geographic region, or that they represent some significant characteristics of the universe (population). In fact, in such cases simple random sampling is inappropriate *because* we know something about the geography or composition of the total population and believe that they are important to the hypotheses or unanswered research questions. In cases of known composition, a stratified random sampling technique is needed because we desire to improve the "representativeness" of our sample by choosing sub-samples that quantitatively reflect important known differences in the make up of the population to be studied.

Stratified Sampling

There are good reasons, when we know something about the population (universe) to be studied, to select a sample that reflects the variability of the population. There are two kinds of stratified samples: proportionate and disproportionate. The idea in proportionate sampling is to guarantee representativeness with regard to the property that is the basis of classification. Disproportionate stratified sampling is efficient when the researcher wishes to compare groups. Examples of each method follow.

Proportionate   Returning to our example of the retail chain that wants to determine "What proportion of rural customers will travel at least twenty miles to purchase hardware products?" let us assume that this chain has secured a list of customers residing in the rural region. The list came from the chain's parent company that has a large credit card clientele in this region. Because of this database, the chain knows something about the demographics of each consumer. With this information it is relatively easy to stratify the customer list on the basis of some characteristic, i.e., to categorize the list by annual income, age, education, family size, or other attribute deemed important. In stratifying the customer list, subpopulations can be identified and samples drawn from each demographic category. Remember that the categorization is driven by theory or hypothesis. In this case, one theory (the basis of the categorization) is that age influences the distance a consumer is willing to travel to purchase hardware. Without detailing the theory, let us assume that elderly consumers will not travel long distances to purchase hardware. Thus, the categorization of age groupings might be consumers under age sixty-five years and those sixty-five and older, or some other carefully considered design for the stratification. In a proportionate stratified sampling technique, the percentage of all customers in a category (or class) is used to determine the percentage of the sample population that will be drawn from that category. This sampling method, then, assures that each category of the determining variable is *proportionately represented* in the sample.

Disproportionate   In other research questions, it may be more important to determine the responses of particular categories more than others. That is, we may seek a *disproportionate* sampling of a particular category or subpopulation. In our case of the retail chain, they probably would value the responses of customers that frequently purchased hardware, rather than those of non-frequent purchasers. Therefore, it would be desirable to stratify the universe in terms of their frequency of purchase of hardware and to select a greater proportion (disproportionate) from the frequent purchasers than from the infrequent buyers. A conscious decision is made in this case to bias the sample toward the responses of a single subpopulation and to weigh less the answers of another group. It also requires important knowledge of the variability of a particular variable or subpopulation.

A disproportionate random sampling strategy is also used when one anticipates that a population segment will be inclined not to reply to a survey. In this situation a sample is drawn from the reluctant group disproportionate to their proportion in the universe. That is, one "over-samples" to ensure an adequate response.

Cluster Area Sampling

Sometimes the physical location of the universe makes random sampling (and its variations—proportionate and disproportionate) inappropriate. If we want to survey in an area that is subdivisible into smaller areal units (for example, a census tract is divisible into blocks or block groups, or cities can be subdivided into neighborhoods), we might employ a sampling technique called cluster area sampling. Useful in sampling large geographic regions, cluster area sampling permits us to extract groups from the universe to ensure a sample that represents the entire universe and avoids the bias of selecting only those people (or other observations) located in a particular region (or regions). In cluster area sampling, each subunit of the geographic region under investigation is placed in a list. The subunits where interviewing will occur are drawn randomly. For each subunit selected, all persons are interviewed. When clusters are selected in a large geographic area, field costs (travel and random interviewing) are lower than for other sampling procedures.

When clusters are selected geographically this technique requires the use of a map. From our previous example, (rural customers purchasing hardware) the store's trade area (all customer locations) would be defined and divided into areas of nearly equal size. Criteria for defining geographic clusters are required. Each customer (or housing structure) must be represented in some form on the map and belong to a group (cluster). A list of addresses or phone numbers for every individual or household is needed in each randomly selected cluster because every individual is contacted for the survey. There are advantages and disadvantages to using this method. One advantage is comprehensive understanding of each selected cluster. A disadvantage is that poor preliminary assignment of individuals to clusters can result in large errors.

Summary

The discussion of sampling in this chapter is brief and introduces only some of the considerations that beginning researchers must consider. In developing a sampling system the researcher must consider: (1) the type of data desired, (2) the degree of accuracy (error range in responses) required versus the cost of securing this accuracy, (3) the importance to the proposed study of an area cross section, (4) the importance of an economic, social, or cultural cross section of the population to the study, and (5) the application of the data obtained. There is no general rule for devising a sample. It varies from one research problem to another. The student researcher who plans to develop sampling procedures is advised to study recent references on the subject.

Survey Design: Questionnaire Construction and Administration

We noted at the beginning of this chapter two important concerns (or determinants of accuracy) of survey research: questionnaire design and administration. All questionnaires have two objectives: obtaining relevant data and acquiring these data in a manner that will ensure the greatest possible degree of accuracy. We also commented on the expense associated with securing accurate data by the survey method.

Survey research is used most often when the data necessary to test a hypothesis do not exist. It is important, therefore, for a researcher to become aware of standard secondary sources of information and their dates of publication. For example, if the researcher seeks information regarding the distance rural customers are willing to travel to purchase hardware in northern Pennsylvania, it is likely that such information is unavailable. On the other hand, it would be unwise for a researcher to conduct a survey in 1994 to learn of the number of retail hardware establishments and level of sales in the counties of rural Pennsylvania because that data is published every five years in the *U.S. Census of Retailing*.

The researcher must also be concerned with the validity of secondary data. One concern is whether or not the information is too dated to be used. For example, if a research problem executed in 1998 requires family income levels at the census track level to test the hypothesis that income level was interrelated to family status, that data would be nine years old (collected in 1989) and outdated. In this case, it would be wise to conduct a survey of families residing in the census tracts in question. In any case of questionnaire construction and administration, rules must be followed to minimize error and to maximize accuracy.

Questionnaire Design/Construction

A questionnaire is an instrument for obtaining specific information relevant to a research study; it is a series of questions that are administered according to a specific plan.

The questionnaire style determines what type of responses the researcher receives. When designing the questionnaire, we must be precise. Often, "yes" or "no" will not provide adequate data and other styles must be developed to secure data appropriate for the analysis. In our case of the retail chain and hardware shoppers, we must decide before we collect the data (design the survey question) whether we wish to calculate some measure of central tendency (mean, median, mode) or variance, or simply report that a certain percent of the sample responded one way, and the remainder answered another way. The researcher could, for example, ask each respondent "How far (in miles) did you travel the last time that you purchased hardware products?" The answer would be open-ended and could range, theoretically, from zero to infinity. The sample data could then be used to calculate the average distance (a measure of central tendency) that rural households of the study area travelled to purchase hardware.

On the other hand, if the researcher asked: "Into which of the following categories do you fall?" and provided this statement, "The last time I purchased hardware products, I travelled: (a) 0–10 miles, (b) 11–15 miles, (c) 16–20 miles, or (d) more than 21 miles." This design yields discrete categories. The data would permit comparisons of subcategories, or cross-classification with other variable categories, such as age groupings. However, categorical data do not permit the calculation of measures of central tendencies. It is important, therefore, to know the nature and type of data desired for analytical purposes to .determine the design of the questionnaire and the response categories.

Space does not permit discussion of all closed response styles, however, the common styles include the checklist (respondents check the answers that apply) and several scales: intensity, frequency, distance, and ranking. Carefully review those appearing in Figure 5.1.

The nature of the research problem will determine whether the responses will be open-ended or closed, as in Figure 5.1. The open or free response format allows for maximum freedom in answering the questions, but tabulation and classification of their responses is imprecise and cumbersome. The closed response form limits the response to specific answers that can be classified and analyzed efficiently and easily. However, it forces the respondent to choose a single answer which may not be a true answer. The open and closed response forms represent the extremes. Many researchers employ questionnaires that combine both open and closed response forms in the same instrument.

In addition to the issue of precision, the researcher must follow well-established rules of questionnaire design and administration. These fall into four general categories: questionnaire and question *length,* question *style and content, organization* of the questionnaire, and *administration* of the survey instrument (see Figure 5.2).

The length of the survey instrument and of each question are important to the success and accuracy of the survey data. The longer the survey, the less likely that the "average" person will participate. Thus, representativeness of the sample can be affected. Further, even when an individual agrees to be interviewed, it is possible that fatigue due to an over-lengthy survey may lead to frivolous or less than thoughtful

FIGURE 5.1

Examples of response styles. After Lounsbury and Aldrich: *Introduction to Geographic Field Methods and Techniques*, Charles E. Merrill Publishing Company, 1979.

CHECKLIST TYPE (variety of possible answers; all those that apply are checked)

During the last five years, has the production of soybeans on your farm decreased because of too little rainfall?

[　] No noticeable decrease during this period.
[　] Slight decrease some years.
[　] Slight decrease every year.
[　] Serious decrease some years.
[　] Serious decrease every year.

INTENSITY SCALE (quantity, intensity, "how much")

Airport noise is a serious health hazard in this area.

[　] Agree strongly with this statement.
[　] Agree moderately with this statement.
[　] Undecided.
[　] Disagree moderately with this statement.
[　] Disagree strongly with this statement.

FREQUENCY SCALE (frequency of an activity, "how often")

During the last year, how often did you visit Black Lake Recreational Park?

[　] At least once a week.
[　] Once a month.
[・] Once every two or three months.
[　] Once or twice a year.
[　] Not at all.

DISTANCE SCALE (distance from one activity to another, "how far")

How far do you travel (one-way trip) to work?

[　] Less than one mile.
[　] One to 5 miles.
[　] 6 to 10 miles.
[　] 11 to 25 miles.
[　] over 25 miles.

RANKING SCALE (priority, "relative importance")

The opening of the Metro Medical Center has had some impact on this area. Which of the following do you consider to be most serious (put "1" in the space next to that item)

[　] Traffic congestion.
[　] Noise.
[　] Shortage of parking areas.
[　] Lowering of property values.
[　] Unsightly structures; detracts from the beauty of the neighborhood.

If you could give **two** answers, which of the items above would you choose second? Put "2" next to that item.

If you could give **three** answers, which of the items above would you choose third? Put "3" next to that item.

responses. Finally, with long survey questions, the respondent is more likely to become confused by the question.

Language is the key to communicating the particular concept or issue for which the researcher seeks a response. Clear language is essential. The words and their meanings must be straightforward and easily understood.

Understanding the audience is crucial. If the researcher seeks information from executives, more sophisticated language is possible. However, the general population contains a great variation in educational experience and cultural background. Accordingly, simple words (three syllables or less) with easily understood meanings are the norms in survey construction.

Additionally, the researcher must avoid ambiguous wording, objectionable phrasing, and leading questions. Ambiguous wording may

FIGURE 5.2

Rules and guidelines for questionnaire design and administration.[5]

1. *LENGTH*

   a)   Keep questionnaire *short* or of *manageable* length.
   b)   Keep each question short. Make it long enough to
        ask the question but avoid long sentences that can
        be ambiguous and confusing.

2. *STYLE/CONTENT OF LANGUAGE*

   a)   Phrase each question carefully.
   b)   Use *easily understood language* that is appropriate
        to your audience (sample).
   c)   Select *words with similar meanings* to everyone.
   d)   Avoid *ambiguity*. Be clear in meaning.
   e)   Avoid *objectionable phrasing*.
   f)   Avoid *leading questions*.
   g)   Limit questions to a *single idea*.
   h)   Provide all possible *alternatives* or *none*.

3. *ORGANIZATION*

   a)   Provide an *accurate context for the survey* and its
        questions.
   b)   Carefully *tie each* survey *question to* a research *goal*.
   c)   Try to organize questions into a natural sequence by
        grouping similar questions.
   d)   Place enjoyable questions first and potentially
        objectionable questions last.
   e)   Limit open format questions and place them strategically.

4. *ADMINISTRATION*

   a)   Provide a *context for a question* when necessary.
   b)   *Protect* the respondent's *feelings*.
   c)   Do not assume factual knowledge by someone else;
        avoid second-hand knowledge.

make the respondent wonder what you mean. Do not ask "When did
you last visit a health practitioner?" when you mean "When did you
last visit a physician?" Don't ask "How much do you earn?" Rather, ask:

---

5. Most of these appear in D. C. Miller, *Handbook of Research Design and Social Measurement* (New York: McKay Co., Inc., 1977).

"Into which of the following income categories do you fall?" Read the categories and ask the respondent to stop you when you read the category into which he or she falls. This is much less objectionable than the previous format. Leading questions give biased answers. Don't ask: "As a patriotic American, do you support defense programs?" Rather, ask "Did you favor Congress's recent vote to fund the XYZ missile program?"

Another rule in questionnaire construction is that each question contain only a single idea to avoid error or difficulty in the interpretation of responses. If a researcher asks: "Do you car pool for work, shopping, or recreation?" then a "yes" response indicates that car pooling may occur for one, two, or all three of these activities. The researcher cannot possibly extract meaningful information from this. To assume that the respondent car pools to all three could lead to erroneous conclusions.

Finally, it is necessary to provide all alternative responses to a question or none. Research has demonstrated that if a partial list of possible responses is given to a question, respondents tend to favor these over reporting open-ended response of their own. The result is inaccurate information.

Organization of the survey questions is important to the success of its execution. It is necessary to provide a context and to be certain that superfluous questions are not included. If a proper context is provided and the researcher ties each question to a research goal, then out-of-context questions can be avoided. The survey should make sense to the respondent, and certainly not be objectionable.

It is important to organize questions into meaningful groups that are related to a single issue or purpose, and to place potentially objectionable questions at the end of the survey. For example, questions related to demographics often appear at the end of a survey.

Finally, with regard to organization, open-ended questions that are essential should be included in the survey. However, they should be limited because they are demanding of the respondent, increase response time to the survey, and are more difficult to analyze.

Administration of the survey has several important elements. The administrator has the responsibility of providing context, whether in writing or verbally, during a personal interview. The administrator must also protect the respondent's feelings during the interview process. Finally, second-hand knowledge must be avoided in order to avoid error. The research design must strictly identify "who" is to be interviewed. During the survey administration phase interviewers must adhere to this design. One cannot assume factual knowledge from a friend or relative who wishes to answer for the sampled respondent. For example, a respondent may know that a spouse has strong feelings about the construction of a trash-burning incinerator. However, that spouse may not be aware of the reasons, or sentiment (*the why*) for such opinion.

Survey Administration Methods

There are four methods of administering questionnaires: the personal interview, where the interviewer and respondent are face-to-face, the mail questionnaire, the group self-administered questionnaire, and the telephone interview. Each method has advantages and limitations.

Although the most costly method in time and money, the personal interview is often the most effective means of acquiring information, both quantitatively and qualitatively. The interviewer should have questions clearly in mind and properly worded so that he or she need not refer to any notes or write down any responses. This is considered desirable because careful note taking, usually valued so highly in research, interferes with candid answers on the part of the interviewee. For best responses, a relaxed, informal, spontaneous atmosphere is desirable. It is sometimes necessary to gather specific data by the interview method where exact questions and note taking are essential, although this is seldom necessary in introductory research problems. For anyone contemplating interviewing, we advise you to consult a more complete work on interviewing.

The mailed questionnaire has fallen into some disrepute in recent years because of the quest for mass information and the availability of the mails. As a result, those who receive such requests tend to disregard them. It is, therefore, all-important to cultivate the receptivity of the respondent in any way feasible. One good list of suggestions includes eleven things the researcher can do to solicit a reply:

1. Ask only for information not available from other sources;
2. Ask only important and significant questions;
3. Try to engender a desire within the respondent to answer questions honestly;
4. Do not ask questions that are to be answered objectively but which imply further explanation;
5. Do not promise the respondent a summary of the questionnaire unless the researcher fully intends to keep that promise;
6. Try to develop a reason for the respondent to answer;
7. Use a questionnaire which is not too lengthy;
8. Encourage the respondent to sign the instrument;
9. Do not use humor or personal reference; and

10. Send an explanatory cover letter with the questionnaire but unattached[6]
11. It is also wise to include a self-addressed, stamped envelope.

The group, self-administered questionnaire is usually completed in group sessions with the interviewer present to provide directions and support. Normally a group of respondents represent a select group—members of an organization, students, or interest groups. The group, self-administered method is limited to studies that do not require a cross section of the study area nor the population.

The telephone interview is expensive and highly flexible. It can be used to obtain additional information in following up a personal interview or to prepare a prospective respondent prior to a person-to-person interview. It also is used when the questionnaire is short and does not deal with sensitive issues. Its major disadvantage for geographic field work is that it is difficult, and often impossible, to match a telephone number to a precise location. Further, the widespread use of the telephone to conduct a wide variety of promotional and advertising campaigns has resulted in a widespread reluctance to cooperate by potential respondents.

Compilation and Tabulation

In geographic research, the data obtained by interviewing must be matched precisely to a location on a map to reveal spatial patterns. Each interview must be numbered and the number located on a base map at the time the interview is given.

The compiling of data from closed response questionnaires is not difficult and may be done by hand or computer. Compilation of the open response form is a great deal more complex. The open response form often results in a wide range of answers, and if the answers are to be classified, a systematic method must be developed to place the diverse responses into general classes.

The acquisition of information not readily observable presents several problems to the researcher. Each research problem is unique, and it is critical to determine precisely what information is essential to a given problem. This determination bears directly on the design and administration method of the questionnaire, as well as sampling procedures used and the way the data are to be tabulated and analyzed. There are several good references on proper questionnaire construction in the area of educational and social research. The student developing such a research tool should read one or more of them carefully (see Appendix A).

---

6. J. Francis Rummel and Wesley C. Ballaine, *Research Methodology* (New York: Harper & Row, Publishers, 1963).

Analysis of Data

.

Among the uses of the scientific method are analysis and prediction. For example, it is at the analysis stage of research that the geographer may wish to weigh, compare, and test his spatial hypotheses for acceptance or rejection. If it is found that the actual location and distribution of the variable under observation corresponds to that stated in the hypothesis, he accepts the hypothesis. The measurement of the extent of the similarity of pattern between the hypothesis and the "real world" must be determined by logical deduction or induction with the aid of statistical and cartographic techniques of analysis. On the basis of such tests, laws or norms may be established—at least a norm for the limited universe enclosed by the bounds of the research project involved.

Logical analysis by means of verbal symbols is a common process in science as well as in everyday reflective thinking. In this process, the procedure is guided by the laws of logic, and the results are expressed in words designed to convey accurate meaning to the recipient. Geographic research has always depended in large measure on the researcher's ability in verbal analysis. It is impossible to imagine a time when this will not be an important part of research work.

Evaluation of Data

The first step in analyzing the collected data consists of organizing, compiling, and comparing one set of data with another, and evaluating. The researcher, in evaluating the acquired data, must make judgments about: (1) their relevancy to his or her particular study, (2) the total quantity, and (3) their quality.

With the statement of his or her problem firmly in mind, he must make judgments for each set of data. This might include a rank-order classification. The amount of data collected must be totaled, and the researcher may find that he or she has accumulated an excessive amount of data concerning a specific aspect of his or her problem, or, on the other hand, he or she lacks sufficient information concerning some part of his or her study. In the latter event, he or she obtains the needed specific data by returning to library sources or to the field. Usually, this requires obtaining a small amount of very specific information dealing with one or two points.

Finally, the data must be evaluated in terms of their quality or accuracy. In the case of published materials, the researcher must consider the source—who wrote the article, how was the information obtained, and on what basis were the conclusions or predictions made. He or she should also note the dates of the articles or books consulted, the kind and amount of illustrative materials, and the bibliographies. In regard to existing aerial·photos and maps, consideration must be given to the extent of the research area covered, the type of phenomena illustrated, the scale, date, and accuracy, that is, how a given map was produced and what sources of information were used in its construction.

The researcher must judge the accuracy of his or her own field work impartially, and he or she must determine its validity for documentation purposes. The evaluation of data most likely will result in the selection of certain data that prove to be essential and basic to the research problem, to the rejection of some information because of some inherent weakness or unsuitability, and to the setting aside temporarily of other data to be reevaluated at a later date.

Cartographic Analysis and Presentation of Data

After the significant data have .been identified and isolated, the researcher can use any of a wide variety of cartographic devices to present the data in a manner that makes analysis more effective. Most of these cartographic procedures involve precise measurements, and they overlap with statistical or quantitative devices. It is not always possible, nor is it necessary, to make sharp distinctions because cartographic analysis supplements statistical analysis and vice versa.

Since geography is the science of spatial analysis, it is likely that maps of some type will be included in any research study. A well-constructed map is a scientific document like a graph, a table, a diagram, or a model. Like these devices, the most appropriate types of maps will vary from one research problem to another.

In general, there are two major ways to present data on maps to preserve the true shape and area of phenomena. One method is the use of dots with numerical values assigned to each dot. Dot maps are commonly used to show population, production of commodities, and other phenomena. The other device is the use of isopleths, or isolines, or isograms. The prefix "iso" means *equal* and describes lines connecting points of equal value. For example, an "isotherm" runs through places of equal temperatures; and "isochrone" expresses equal amounts of time such as might be involved in travel or growth processes. "Isopleths" may be used, not only to show actual existing phenomena, but also tendencies. An "isallobar" connects places that have the same tendency of air pressure change within a specified period of time.

The researcher must decide what type of cartographic device is most significant to his or her particular study. Maps can be constructed to show the spatial distribution of qualitative or quantitative (or both) aspects of phenomena. They may be single-featured, showing the distribution of one datum, multiple-featured showing the spatial distribution of several data, or they may show the ratio or relationship between two or more data. For example, a map might be constructed to show by dots the location of all aluminum-producing plants in the United States. This type of map would show the distribution of only one item of qualitative information. For purposes of geographic analysis, the same map would be improved tremendously if it were to include quantitative information indicating "how much" was produced at each plant by the assignment of appropriate numerical values to each dot or to each dot of a different size. Such a map shows two related data and illustrates important, although general, information about the aluminum industry. If the researcher were to refine the map further by plotting the sources and amount of bauxite (aluminum ore) and by adding isolines indicating power costs, it would be discovered that the bauxite sources and the aluminum plants have little relationship spatially, but that power costs appear to be most significant since all aluminum plants are located in very low power cost areas. Thus, a relationship has been unearthed, one that may be further investigated and tested by using statistical techniques discussed later in this chapter.

Ratio maps are usually significant to geographic studies. They require previous computation of data because they show relationships of amount to area, percentage of one or more items to the total, or some other statistical relationship between two or more phenomena. Simple examples are the widely-used population density maps (*number* of persons per given unit of *area*), or effective growing season maps (80 *percent* of the *average* frost-free season).

Some ratio maps may be quite complex. As a hypothetical example, let us assume that in a given subhumid area, the amount of precipitation is a critical factor in determining whether or not it will be economically feasible to attempt to grow a given crop. Let us also assume that this crop needs nine inches of surplus rainfall (beyond the evaporation rate) during the growing season (May–September), and that the area receives eight to ten inches depending upon the locality. Let us assume further that data indicating total monthly rainfall and evaporation are available for various locations. The question might be, "What are the chances for success in any given year at locations A, B, C?" This may be shown by constructing a map with several isolines, each representing a given percentage of years in which nine inches or more of surplus water may be expected. Such a map requires several manipulations of data: the average monthly evaporation rates subtracted from the average monthly rainfall rates for each location during the growing season, computing the percentage of years that each location will receive at least nine inches of surplus water during the growing season, and plotting several isolines (perhaps the 80 percent, 90 percent, etc.) on the map.

The researcher faces two pitfalls in constructing cartographic devices for analysis. There has never been a map constructed of the world, or of a large segment of the earth, that is not distorted by some degree. It is impossible to represent a curved surface accurately on a flat plane; and small scale projections may show either true shape or true area or, perhaps, neither. It is not likely that the student researcher who is engaged in his or her first research problem will be dealing with a matrix so large that the earth's curvature will be significant. However, in the event that he or she does deal with spatial distributions on a world or a continental scale, care must be taken to select the appropriate projection that illustrates his or her data without distortion. The second danger is inherent to all map construction. Values must be assigned for any dots, lines, or legends that are used. Extreme care must be taken and a well-defined rationale stated for assigning numerical values or legend classification to ascertain that relevant spatial patterns are not obscured or that actual misrepresentation results.

To illustrate this point with very simple examples, let us assume that a researcher is constructing a contour map of an area (lines connecting points of equal elevations) and that he or she decides to use twenty-foot contour intervals. If all the area is under twenty feet elevation except for one small hill summit, the map will show only the outline of this one small area, and the rest will appear to be flat. In reality, the area may be a "badlands" with nearly perpendicular slopes and sharp gullies between zero and nineteen feet in elevation.

Let us assume further that a researcher constructs a map of an area to illustrate the percentage of the population that has blue eyes. He or she elects to show this spatial distribution by devising a legend with categories such as 91 percent and above, 81 to 90 percent, and so forth. Perhaps the map might show a small island in the 91 percent-plus category surrounded by a very large area in the 81 to 90 percent category. In the latter area, perhaps 88.7 percent of the population is blue-eyed. By modifying legend values downward by only two percentage points, the researcher has changed the spatial distribution drastically. A general rule to follow in assigning values to legends, isolines, dots, and the like is to ask the question, "Will the proposed value categories show the spatial distribution of the significant aspects of the data as they actually exist?"

Statistical Analysis and Presentation of Data

In data presentation, the most used expressions of measurement are those of descriptive statistics, such as the mean, median, mode, standard deviation, locational quotient, variance, quartiles, and similar tools. Bar graphs, flow charts, scattergrams, pie charts, and a number of related tools may be developed using statistical procedures, but may be presented cartographically. In organizing and analyzing data, the importance lies in the accuracy of the operation and its appropriateness for the task, not in whether it is part of a particular technique.

Special techniques of spatial analysis involving more sophisticated uses of statistical procedures have been developed recently. This development has been spurred by new techniques of inferential statistics, by the creation and widespread use of the electronic computer, and by the evolution of geography into a more exact scientific discipline. Because science seeks to be as precise as possible, geography has accepted the need for more exact measurements. This need is as great for the normative sciences as it is for the experimental sciences. Numerical statements are more precise than verbal expressions. They are also more universally understood.

It is well to remember that geography is not an experimental science. Geographic analysis does not contemplate the effect of controlled variables, but it observes the locational pattern as it is found and attempts to account for its distribution. Since the number of independent variables affecting any distribution is large and not experimentally controllable, it is necessary to accept the philosophy of *improving* the accuracy of the analysis as a major goal of research geography. Total explanation may be ideally desired, but often it is impossible, or at least too time-consuming and expensive to be feasible. The large number of uncontrollable variables makes it advisable to employ some statistical method of identifying the effect of the variables under consideration. Thus, their separate effect on the distribution, as well as the total amount of explanation accomplished by the study, can be measured with some accuracy.

At the present time, the average geography student has not been exposed to statistical procedures to the extent that he or she has been trained in verbal expression. It has been only recently that departments of geography have begun to require courses in quantitative techniques. It may be assumed that, as the student advances in professional training, he or she will be exposed to increasingly more sophisticated statistical techniques. However, for the beginning researcher, a working knowledge of a few statistical techniques will, with the help of a basic statistical manual, make it possible for students to understand articles in geographic publications and to engage in research studies.

Three of the most commonly used statistical procedures in geographic literature are scatter diagrams, regression analysis, and coefficients of correlation. With a working knowledge of these, plus some elementary model concepts, the undergraduate and beginning graduate student is capable of understanding most of the research reports published and to perform meaningful geographic research independently. The full extent of the researcher's specific needs can be determined only by the individual and the advisor in light of the research problem faced. The skill of analysis cannot be measured in terms of the number of techniques and tools used. The purpose of research is to answer the questions asked. The selection and use of the proper procedures are only part of the problem-solving process. With this in mind, let us consider the three most commonly used statistical devices in geographic research.

The Scatter Diagram

The relationship among areally associated phenomena is usually not apparent until some device for detection is applied to the data under observation. One such device is the scatter diagram, or scattergram. It is easy to construct, uncomplicated in nature, and easily interpreted. It not only visually demonstrates the general character of the relationships, but it directs attention to the cases that do not conform to the pattern.

The scattergram is a device for visually showing the relationship between two variables for a selected number of observation units. One variable is measured on a horizontal axis and the other on a vertical axis. Where the two lines intersect, a dot is made indicating the location of that observation unit on the diagram. For example, the average maximum summer temperature (June, July, and August) for Phoenix, Arizona, is 102°F., and the elevation is 1,083 feet. Phoenix is repre-

FIGURE 6.1

Scatter diagram showing average maximum summer temperatures and elevations for selected stations in Arizona.

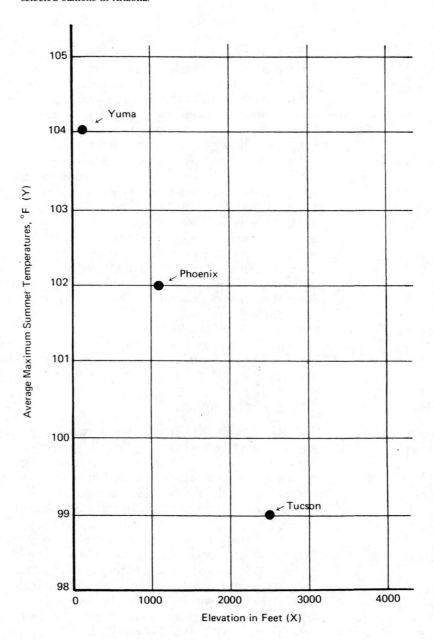

sented on a scattergram as a point, as shown in Figure 6.1. Yuma at 104°F. and 138 feet in elevation is located higher on the temperature axis but lower (to the left of Phoenix) on the elevation axis. Tucson at 2,410 feet above sea level is to the right of Phoenix on the "X" axis, and its summer maximum temperature (99°F.) is lower on the vertical scale.

There are three questions of importance to note in reading a scattergram: (1) Do the points tend to be arranged in a sloping straight line? (2) Does the line slope from upper left to lower right, or from lower left to upper right? and (3) Do the dots arrange themselves close to the line of slope? Each of the answers reveals something about the nature of the data and directs future research. The three questions should be considered separately.

If the dots are arranged in a sloping line, a relationship between the two variables is indicated. If the dots are widely scattered, are circular in arrangement, or are in a straight horizontal or vertical line, no relationship exists.

Since the scale of the vertical axis (Y) usually progresses from bottom to top, and the horizontal scale (X) runs from left to right, the slope of the line gives additional information. If it slopes from lower left to upper right, this means that, as the magnitude of one value increases, so does the other. This is called a *positive, or direct* correlation. However, if the line slopes downward from the upper left corner to the lower right, it shows a *negative, or indirect* relationship between variables. This is the case in Figure 6.1, where Yuma with a low elevation has a high temperature, while Phoenix and Tucson have higher elevations with lower temperatures. Of course, any scattergram used in a research project will have more than three observation units, but those in Figure 6.1 illustrate the inverse correlation.

The third characteristic to be noted is the "closeness of fit," that is, how close the dots are to a straight line. As mentioned previously, a scattering of dots shows no relationship. The nearer the dots tend to be to a straight line, the higher the correlation between variables. The exact degree of this "fit" is called a coefficient of correlation. It may range from no relation (zero) to a perfect or 1.00 correlation; and it may be either positive (+) or negative (−), depending on whether the slope tends to the left or right.

Another aspect of a "close fit" should be mentioned at this time. Correlation statistics are based on the assumption of a linear relationship between variables. If the dots tend to follow a straight line with little scatter, the researcher has good reason to assume that there is a linear relationship.

As we shall see later, the scattergram is closely related to the processes of correlation and regression. These three tools are all useful in description, analysis, and prediction. The scattergram, moreover, is a very useful device during the search for hypotheses that might explain the relationship between variables. The scattergram gives the researcher a quick idea of the degree of relationship. If the observations cluster along a sloping line, further investigation is suggested. At the same time, the scattergram calls attention to the isolated case (residual). The residual may become the subject of a separate case study. More often, its value is to direct attention to the presence of a strong independent variable which, when considered in the analysis, will not only explain the deviation of the residual but will improve the fit of all other observation units in the study.

For example, let us assume a hypothetical case where the researcher attempts to develop a system for classifying villages between 3,000 and 5,000 population. He or she hopes to classify them on the basis of occupational structure; and has theorized that manufacturing and selected service occupations (persons employed in grocery stores, service stations, barbershops, etc.) will prove meaningful. The researcher plots on the scattergram (Figure 6.2) the percentages of the total population employed in manufacturing and service occupations for each of the villages in his or her research area. The clustering of dots in distinct groups indicates that there is some basis for the proposed classification. The information brought to light by the scatter diagram enables the classification of villages tentatively as follows:

Group A—"Manufacturing Villages," which the researcher tentatively defines as those with 25 percent or more of the population employed in manufacturing and no more than 15 percent in service occupations. Perhaps this confirms an earlier hypothesis that villages of this type are located near large manufacturing plants on the periphery of large cities.

Group B— "Service Centers," which serve as central places for adjacent rural areas (tentative definitions can be stated as in Group A villages). Perhaps this confirms a hypothesis that villages of this type are located some distance from larger towns or cities.

Group C—"Diversified Villages," which have a wider range of functions and may represent a broader or balanced occupational structure.

Group D—"Strange Cases," where manufacturing and service occupations are relatively unimportant.

The researcher now finds that the "Strange Cases" require further thought, and hypothesizes that these may be farming communities, mining towns, or settlements inhabited by persons employed in commercial fishing or in some other occupations that he or she has not considered. Perhaps they may even represent a resort or a retirement community.

FIGURE 6.2

Scattergram constructed for classifying villages of 3,000 to 5,000 population on the basis of occupational structure.

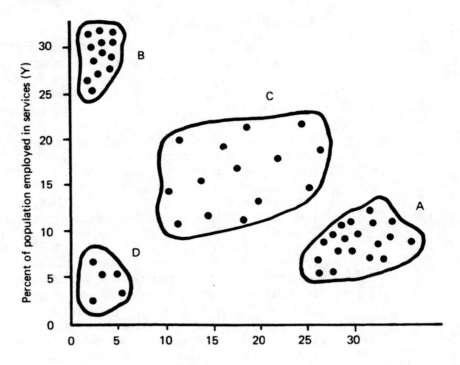

The researcher must now investigate further if he or she wishes to subdivide the "Strange Cases" grouping into meaningful categories.

This hypothetical example illustrates a way in which scatter diagrams may be used to analyze data. In this case, *one* diagram has provided the researcher with information that may serve as the basic framework to classify villages, and it also focuses the researcher's future efforts toward refining his or her classification (perhaps percentage of gainfully employed rather than total population would be a possibility). The scattergram also directs the search for other variables to understand and explain the "no fit" cases.

In the event that such a clustering of deviant cases is found to be concentrated within a limited area, there is sound basis for defining a new region or for recognizing a subregion. For example, northern Missouri has a section referred to as "Little Dixie" that possesses unique voting characteristics. In the regionalizing of the political patterns, this section commonly appears as a distinctive subregion. In accounting for the area, the independent variable of southeast American immigrants and their cultural view would have to be considered.

Regression Analysis

A regression analysis is one method of measuring the relationship between two or more variables. If only two variables are considered, it is called simple regression as contrasted with multiple regression when three or more variables are considered. Regression analysis is a precise mathematical expression of the observation units plotted on a scattergram. It defines the direction and degree of slope of a line drawn closest to the average of all the data on the scattergram. This line is called the "line of best fit" or the "regression line" and is designated here by the symbols "Yc." The equation for "Yc" is $Yc = a + bX$ for simple regression. In the case of multiple analysis, additional "bX" values are designated, as $Yc = a + b_1X_1 + b_2X_2 + \ldots b_iX_i$. Often the trend of the observations on the scattergrams is such that a curved line of regression would come closer to every observation than a straight line. When such a regression line is computed it is termed a curvilinear regression line. The formula for such a regression computation can be found in most statistic texts, and the principle involved is the same as in linear regression. It is more difficult and time-consuming, however, and is usually not necessary for the beginning researcher. For the same reason we shall limit this discussion to the analysis of only two variables. This is called simple regression. In situations where more than two variables are involved in regression analysis the term is multiple regression. In multiple regression the equation becomes $Yc = a + b_1X_1 + b_2X_2 + \ldots b_iX_i$. When working with multiple variables and curvilinear relationships it is advisable to make use of computer programs. Simple straight line relationships, however, may be readily computed by hand. Their application by the beginning researcher gives satisfactory results and furnishes the necessary insights into their value as a quantitative technique. Discussion here will be limited to simple regression, but the same procedure may be applied to multiple regression analysis.

Two characteristics of the regression line are necessary for its description. First is the elevation of the line as established by where it is started on the vertical (Y) axis. This value is called the "Y intercept" and is designated by a small "a" (see Figure 6.2). The second characteristic is the slope of the line or coefficient of regression, designated as "b." ("X" is the magnitude of the variable measured on the horizontal axis.) The actual procedure for finding the value of "a" and "b" is discussed at the end of correlation methods in Appendix F. Note that the two characteristics are independent of each other, but that both are needed to establish the position of the regression line.

The "a" value does not have particular significance in an individual analysis beyond giving the line its starting point. It may, however, be of great value in future studies that use the regression analysis as resource material.

The regression coefficient ("b") shows the relationship between the variables. It expresses the change in the "Y" variable as it varies in relation to the "X" variable. The "b" value answers precisely some of the questions previously raised concerning scattergrams. The direction of slope is indicated by the sign preceding "b," and the incline of the slopes is indicated by the number. A minus sign means that an inverse relationship exists, with a downward sloping line from upper left to lower right.

Formulae for computing the regression line may be found in Appendix F.

One question cannot be answered by "a" or "b." That is, how close are the observations to the regression line? Such a measure of the variation of the dots from the regression line may be obtained by computing the standard error of the estimate. It simply indicates how much of the variation of "Y" is not explained by the changing value of "X." It is computed by:

$$S = \sqrt{\left(\frac{Sum\ of\ Squared\ Variation\ Regression}{Number\ of\ Observations\ -\ 2} \right)}$$

The standard error of the estimate indicates, in absolute terms, on the average how much the dependent observations are to be expected to vary from the regression line. Thus the geographer, knowing the analysis is not perfect, is informed of the amount of variation to be normally expected. Armed with this knowledge, the researcher can ignore cases that are probably insignificant variations and can direct analysis to truly exceptional cases when they occur. A more useful expression of the closeness of fit may be found in the coefficient of determination. This is an extremely valuable tool for analyzing spatial relationships and will be discussed in detail later in this chapter.

The method of fitting a regression line to a distribution of matched variables when "a" and "b" are known is demonstrated by the following example using actual temperature and elevation data for six locations. The elevation in feet of the locations is designated as "X," and the average summer maximum temperature in degrees of Fahrenheit as "Y." The first step is to prepare a table of values (see Table 6.1).

From the table, the researcher may construct a scattergram, plotting the data in order to analyze the distributions better (see Figure 6.3). This procedure, although not absolutely necessary, is advisable in order to fit a line of regression to the data. (In multiple regression problems, it is not possible to show the variables on a scattergram.)

TABLE 6.1

Elevations and Temperatures for Selected Stations in Arizona

Station	(Elevation) X	(Degrees F) Y
1	6,903	78
2	1,083	102
3	5,219	89
4	6,964	81
5	2,410	99
6	138	104
	$\overline{X} = 3,786$	$\overline{Y} = 92$

The regression line, or line of best fit, is drawn on the basis of all paired observation units and is the closest straight line to each and every dot representing an observation unit. The determination of this line is demonstrated in Appendix F. Using the data in Figure 6.3, the line is based on a = 106 and b = −3.68. It is drawn on a diagram by selecting two points on the "X" axis, multiplying the value of each by −3.68, and adding the product to 106. Both of these points are then plotted on the scattergram, and a straight line is drawn which passes through the points. A good test of its accuracy is to determine if it also passes through the mean of "X" and "Y." If it does not, some error has been made.

The regression line's relationship to temperature variations is more meaningful than an average line $(\overline{Y})$ which, in this case, is 92° (it would be shown as a straight horizontal line). An estimate of the temperature of Phoenix on the basis of the average would be 10°F in error. By using the elevation variable in a regression analysis, a temperature of 102.12°F (a + bX) would be predicted. Since Phoenix actually has 102°F, the analysis has failed to explain 0.12°, but it has "explained" 9.88° of the error as estimated by the average. It is obvious that a great deal of the temperature variation from the average can be accounted for by the variation in the elevation of the station.

The distance of the observation dot from the regression line compared to its distance from 92° $(\overline{Y})$ indicates that amount of "Y" variation explained by regression. Some stations are not completely explained by elevation, and some are over-explained. Other variables besides elevation (such as local atmospheric conditions, latitude, and exposure to air masses) affect the temperature, and a more complete explanation could be obtained by a multiple regression analysis. The need for selecting additional variables depends upon the nature of the specific problem.

FIGURE 6.3

Elevations and average summer maximum temperatures for six selected stations in Arizona. Source of data: TABLE 6.1.

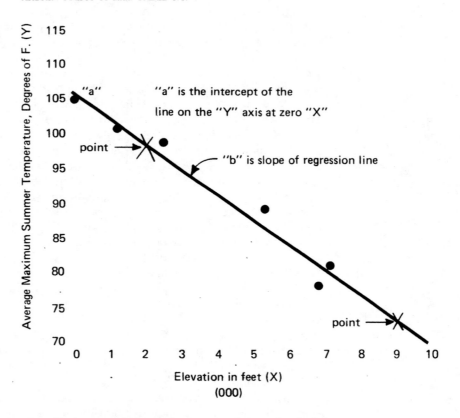

Coefficients of Correlation and Determination

Two useful and frequently encountered statistics in geographic analysis are the coefficients of determination ("r^2") and its square root ("r"), the coefficient of correlation. In the preceding discussion of regression analysis, the relationship of variables was described about direction and slope of the regression line by the formula, $Yc = a + bX$. Furthermore, it revealed that some measure of scatter, that is, how much error is contained in the analysis, could be stated by the "standard error of the estimate" statistic. A more useful measure is the coefficient of correlation and its related statistic, the coefficient of determination.

The coefficient of determination ("r^2") is a measure of the relative relationship between the X and Y variables. In the regression analysis of temperature and elevation (Figure 6.3), a certain amount of unexplained variation in the distribution of temperature (Y) remained after the regression analysis. Stated another way, the total variation in the dependent variable (Table 6.1) ranges from 78° to 105°. For all stations the best estimate of the temperature is the mean of 92°, therefore the best prediction of any unknown station taken from this population would be 92°. The sum of all the individual variations from the mean of 92 constitutes the total, or 100%, variation of this group. By selecting the independent variable of elevation to explain some of this variation we have improved our ability to predict an unknown station from this set immensely. For example, if we were asked to predict the temperature for a station whose elevation is 9,800 feet we would predict approximately 70°F (read from Figure 6.3), rather than the 92° which was our best estimate based on temperatures alone. We have obviously "explained" much of the total variation of temperatures in our study area, but how much? The answer to that question can be found in the coefficient of determination. Theoretically, all spatial variation could be accounted for, but in actual research, geographers rarely provide complete or total explanation. The coefficient of determination (r^2) is a measure of what percent of the total variation in "Y" is explained statistically (or determined) by the distribution of "X." Referring again to Figure 6.3, Phoenix with its 102°F average summer maximum temperature is 10° above the average of 92°F. Regression analysis accounted for 9.88°F of that variation, or 98.8 percent of the total variation of the Phoenix temperature from the average. In this single case, the coefficient of determination is .988, i.e., r^2 = .988. Of course, no researcher would cite an "r^2" for one observation because of the unreliability of such a figure. But this does illustrate how the coefficient of determination is a measure of how much of the variation is explained by the regression analysis.

In Figure 6.3, every observation is imperfectly explained. In other words, the line would not necessarily pass through any dot on the diagram. The equation clearly explains only part of the total, but how much? In this case, an "r^2" for all the observations is .994, which means that for these six stations in Arizona, elevation accounts for 99.4 percent of all variations in temperature. Other variables must explain the remaining 0.006. If some other significant variable were combined in the analysis to arrive at a multiple coefficient of determination (R^2), the resulting figure would range somewhere above 0.944; but it is unlikely that it would ever reach 1.00 or 100 percent explanation of the distribution.

There are various statistical correlations for arriving at coefficients of determination. Some are more precise than others, but all essentially define covariation. For the purposes of the beginning researcher, three of these correlations are discussed in the order of their ease of computation. In the cases of these three correlations, the more easily the coefficient can be computed, the less precise is its measurement.

Tetrachoric Correlation

The tetrachoric correlation is the easiest to compute. It is useful for rapid calculations in preliminary research and for class report preparation. It is also occasionally found in published research reports. It does lack precision, but its results are usually close to those of more complex methods. The correlation is computed by ranking the variables and counting the number of observations in the top half of the array (above the median) for both variables. This number is then divided by the total number of observations to obtain a percentage value. For example, if ten observations were ranked from 10 to 1 in value, and four of them were in the top half of *both* "X" and "Y" (see Table 6.2), the value would be 40 percent. This value corresponds to a coefficient of correlation (r) of 0.81 and may be read directly from a table of percentage values and correlation equivalents in Appendix F.

TABLE 6.2

Hypothetical Data Illustrating a Tetrachoric Correlation

Observation	Rank of "X" Variable	Rank of "Y" Variable	
A	1	1	40 percent of the
B	2 } top	2 } top	observations rank
C	3	3	above the median in
D	4 } 40%	4 } 40%	*both* variables
E	5	10	(A, B., C, and D).
F	6	5	Read Coefficient of
G	7	6	Correlation in
H	8	7	Appendix F.
I	9	8	
J	10	9	

The temperature and elevation data from Table 6.1 may be used to demonstrate the computation of a negative tetrachoric correlation.

In Table 6.3, there are no observations in the top half of both "X" and "Y". The percentage, therefore, would be zero, and the coefficient of correlation by this method is −1.00, a perfect negative correlation. The coefficient of determination (r^2) is 1.00. The tetrachoric correlation indicates perfect prediction, and although it is very high it is not perfect, as may be seen by the scattergram in Figure 6.1.

A more precise method of correlation will give a more accurate correlation value.

TABLE 6.3

Elevations and Temperatures for Selected Stations in Arizona Illustrating a Negative
Tetrachoric Correlation

Station	Elevation "X"	Rank of "X" Variable	Temperature "Y"	Rank of "Y" Variable
1	6,903	5	78	1
2	1,083	2	102	5
3	5,219	4	89	3
4	6,964	6	81	2
5	2,410	3	99	4
6	138	1	104	6

Spearman's Rank-Order Correlation

A second method of correlation is the rank-order correlation. It is also
simple to compute and is frequently used and has 92 percent of the
power efficiency of the Pearson Product Moment Correlation. The formula for its operation is:

$$r_s = 1 - \frac{6 \Sigma d^2}{N^3 - N}$$

where "N" is the number of observations, and "Σd^2" is the sum of the
squared rank differences. Again using the temperature-elevation data,
the correlation is obtained as follows:

TABLE 6.4

Elevations and Temperatures in Rank Order for Selected Stations in Arizona

Station	"X" Rank	"Y" Rank	d	d^2
1	5	1	4	16
2	2	5	3	9
3	4	3	1	1
4	6	2	4	16
5	3	4	1	1
6	1	6	5	25

$$\Sigma = 68$$

$$1 - \frac{(6)(68)}{(6 \cdot 6 \cdot 6) - 6} = 1 - \frac{408}{210} = 1 - 1.94 = -9.4$$

$$r = -.94 \quad r^2 = 0.88$$

By the rank-order method of correlation, the portion of the distribution of temperatures explained by elevation is 0.88. This differs somewhat from, and is a more precise measure than, the 1.00 coefficient obtained by the tetrachoric method.

In order to gain experience in the use of rank-order correlation, let us consider another problem that tests the hypothesis that elevation affects temperature range.

The correlation indicates that .92 of the variation is explained by elevation.

TABLE 6.5

Elevations and Temperatures for Selected Stations in Europe

Station	(Meters) Elevation	(°C) Temp. Range	"X"	"Y"	d	d^2
Bozen	290	22.5	1	7	6	36
Buxen	580	21.9	2	6	4	16
Innsbruck	600	21.1	3	5	2	4
Sterzing	1,000	20.7	4	4	0	0
Schafberg	1,780	14.6	5	2	3	9
St. Bernard	2,470	15.2	6	3	3	9
Sonneberg	3,105	14.2	7	1	6	36

$$\Sigma = 110$$

$$1 - \frac{(6)\,(110)}{(7 \cdot 7 \cdot 7)} = 1 - \frac{660}{336} = 1 - 1.96 = -.96$$

$$r = -.96 \qquad r^2 = 0.92$$

Pearson Product Moment Correlation

The method used to measure most accurately the coefficient of determination at 0.994 for the six weather stations in Table 6.1 is the Pearson Product Moment Coefficient of Correlation. This is the most widely used and best known correlation tool in geographic research. Its computation is comparatively slow and requires careful preparation. As a result, it is somewhat difficult to use. Consideration here of a few points regarding its use and value should render it useful to the beginning student.

The Pearson Product Moment Correlation method is so commonly used that almost every academic institution has it already programmed for the electronic computer. These "call programs" have been designed by the computer manufacturers, and the researcher has only to supply the data in the proper form. The computer is especially needed for working on multiple correlations, where more than one independent variable is used.

When only one independent variable is used, the process is called simple correlation. Its computation is simple when compared to multiple correlation. However, even this computation is difficult to perform unless careful attention is given to each step. The beginning research student will want to master the simple correlations for several reasons. First, because two variables can be shown on a scattergram, the student researcher can actually see the unaccountable variation by the scatter of the dots from a line. The student may compare this visual image with the measurement of correlation and develop an understanding of what a correlation means. Second, simple correlations can usually be computed more rapidly by hand than by computer because most computer centers are too busy for the student to gain quick access to their facilities. In fact, some computer centers consider simple regression problems too easy to justify their use unless a great number of variables are involved. Finally, the method is not so difficult that it cannot be used by any careful worker who is able to add, subtract, multiply, and divide without error.

In order to demonstrate how the Pearson Product Moment Coefficient of Correlation is computed, a simple problem of spatial relationship is worked step by step in Appendix F.

The interpretation of a particular coefficient of correlation is more useful when it is subjected to a test for significance (see Appendix F). There are various tests for significance that tell the researcher the chance that a coefficient as high as the value he or she obtained could be obtained by the accident of sampling. Among such tests, the Chi Square, "t," and "F" are most common. If a value is significant at the .01 level of confidence, it is usually considered "highly significant" because the odds are only one in one hundred that a relationship as high as the one found could be obtained by accident. Another commonly cited level of confidence is the .05 level, referred to as "significant."

While the significance of any particular coefficient of correlation is directly related to the number of observations in the sample considered, some general values may be placed on various coefficient levels. In general, coefficients below .25 are not indicative of a meaningful relationship between or among variables. Between .25 and .50 suggests a small to moderate relationship, while .50 to .75 may be considered substantial. Above .75 is considered high, or very high if over .90.[1]

---

1. J. P. Guilford, *Fundamental Statistics in Psychology and Education* (New York: McGraw-Hill Book Co., 1965), p. 145.

Models

The construction and use of models by the researcher has long been an important aspect of geographic research. Some cartographic models, such as aerial photographs (iconic models) and maps (analog models), have altered the scale or substituted a property in order to study reality better. In recent years, emphasis has been directed toward the use of symbolic models for analysis, especially mathematical symbols to define the parts of the model, and graphic models of various types, especially those dealing with space-function relationships.

The beginning researcher should know how to construct and use symbolic as well as more standard types of models. A symbolic model is a description of a set of phenomena stated in a way that allows any change in the various parts to be measured, and its effect on the whole to be predicted. These models focus on a set of theoretical phenomena and may or may not reflect reality. Additional variables, however, may be added to a model until reality is approached; but the basic model is designed to consider the interrelationship of a selected set of variables. Consider, for example, one of the simpler models—the regression equation for predicting temperature change. If the average lapse rate of 3.5°F per 1,000 feet is expressed as $[-3.5° F (X)]$, then any change in either the elevation or the rate of temperature variation can be assessed. It is evident that, in reality, many other variables affect temperature change. A nonscientific thinker whose knowledge of the "reality" of temperature change is limited to the use of the home refrigerator might depreciate the value of this model that predicts temperature decrease in relation to increased elevation. However, for the climatologist and the geographer, such a predictive model has proved to be the basis of a major concept concerning atmospheric processes.

Because of the many types and subtypes of models, no attempt at a comprehensive discussion is made here. A few examples will serve to illustrate their usefulness. A graphic space model depicting the incidence of serious crimes for a major city as they are spatially related to the home of the criminal has been constructed (see Figure 6.4). This is a single variable space model, and its use is mainly as a hypothesis to serve as a way to observe the distance-function elements of a spatial problem. In itself, the space model is not intended to be a solution to a problem. Such models may become extremely complicated as additional variables with varying degrees are added. It is apparent that this complication would soon limit its usefulness and that some form of mathematical definition of the parts would be more manageable.

Another example of one of the more common mathematical models is the gravity, or interaction, model mentioned in Chapter 2. Based on the law of gravity, such models define spatial interaction. The most elementary type treats only two variables: population and distance. The

FIGURE 6.4

Crime-Distance Model. Domicile of the criminal is at the center of the diagram. Each dot indicates where one percent of the total crime is committed in relation to the criminal's domicile. Lines denoting zones are representative of one-half mile intervals.

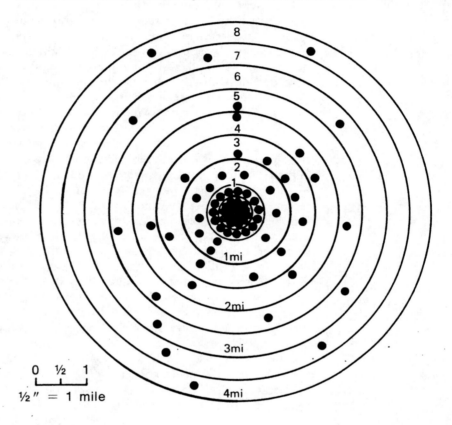

interaction between the points is predicted as equal to the product of the population of the two cities divided by the distance between them:

$$i = \frac{P_1 \ P_2}{d}$$

The resulting figure is an index of the interaction that may be compared with a similar index figure for other cities. To illustrate, let us consider a model based on the data shown in Table 6.6.

TABLE 6.6

Populations and Distances between Four Hypothetical Cities

City	Population (000)	Distance in (00) Miles			
		Metropole	Suburbia	Urbann	Bigtown
Metropole	10	0	4	6	7
Suburbia	20	4	0	5	8
Urbann	30	6	5	0	1
Bigtown	40	7	8	1	0

If there were 100,000 automobiles traveling between Metropole and Urbann, how many would be expected to travel between Suburbia and Bigtown? Using the mathematical interaction model, the following results are obtained:

Case	Cities	(000) Population × Population		= Total	÷ Distance	= Index
A	Metropole Urbann	10	30	300	6	50
B	Suburbia Bigtown	20	40	800	8	100

It is readily apparent from the computation that for every 100 automobiles traveling between Suburbia and Bigtown, there are only 50 cars traveling between Metropole and Urbann. This is a ratio of 2:1. Therefore, 200,000 autos would be expected to travel between Suburbia and Bigtown, since 100,000 traverse the Metropole-Urbann route.

As different phenomena are considered, modifications must be made in the model to better define the interaction. For example, the interaction rate of airplanes between two points would differ from ground retail trade traffic. Therefore, constants and other variations in the formulae are needed to better predict the interaction. Such models can predict only on the basis of the variables employed. In reality, the element of accident, or chance, may partially defeat the most carefully drawn conclusions. Fortunately, if more "reality" is desired in model prediction, there are models that consider the element of chance. These are called "stochastic" models. They add an ingredient of randomness to the structure of the model. This is a partial answer to the criticism that models do not sufficiently resemble reality. By incorporation of the chance factor, the stochastic model predicts what might happen and what the probability of its occurring is.

The Electronic Computer

The electronic computer has been an essential tool in research for decades. In fact, the computer and the "Quantitative Revolution" in geography developed simultaneously. Today, most departments of geography have computer equipment and offer courses that involve the research student in some phase of computer usage.

The computer is a highly versatile and specialized implement. As with any specialty, a considerable amount of language (jargon) is associated with its parts and processes. The researcher must become familiar with the language to some degree, and must have some knowledge of the computer's capabilities. He or she must not, however, become involved with the machinery to the extent that he becomes a technician and loses his identity as a geographer.

The computer, though useful, is not an end in itself. It cannot do anything that a person cannot do; but it can do it quicker, more accurately, and without fatigue. For example, it can add several hundred thousand 16–digit numbers in a fraction of a second. It can continue such a task indefinitely, without tiring and without error; and it will store in its memory any or all of the information for instant recall at any future time. Such a machine lengthens the working life of the researcher and is to be highly valued.

Today, it is the rare geographic research project that does not involve computer usage in some aspect of the research work. Portable microcomputers may be used in the acquisition of field data in some cases, and computers are commonly used in the analysis and presentation of research data. In some research, the computer is used to a degree that virtually excludes other forms of analysis. It has become a commonplace tool to present research data cartographically. Its use allows greater sophistication in measuring and map sampling than conventional cartographic methods, and frequently is less expensive. The widespread use of the word processor has made the writing of research reports, theses, and articles faster and more efficient.

Computer scientists refer to the work their machines perform as *information processing*. Information is transferred to the machine as *input* and the results of the operation are referred to as the *output*. The input may be in the form of teletype terminals, tapes, magnetic tapes, floppy disks, or a number of other methods. Specialists who are trained to communicate with the computer are called *programmers,* and they prepare the *programs* or *software,* that is, the instructions for what the computer should do with the data. Such instructions are written in a special *automatic programming language* such as Fortran, Cobol, C, or some other language, and are translated by the computer into *machine language.* A typical input, for example, on a floppy or hard disk contains numerical information specially prepared for the project, and is often *canned,* that is, prepared and kept stored in program libraries to be used when needed.

Great strides are being made in the development of both hardware and software at the present time. Each generation of computers is faster, more efficient, and can perform more functions. Equipment used today could very well be obsolete or dated in a short period of a few years. For example, slightly more than a decade ago, most computer mapping could only be accomplished on a mainframe, or very large, computer. Today, geographers literally carry numerous maps of cities and regions in the memories of their portable computers. The research geographer of the future will rely heavily upon the computer to store information, manipulate data, generate new information, perform complex calculations, and prepare maps.

Those student researchers with little or no research experience are often faced with the question of how much and when in the research process should the computer be used. There is no one ironclad rule or guideline to follow because each research project has its unique characteristics. Basically, computer usage depends on (1) the nature of the research problem and the type of data to be collected, analyzed, and presented, (2) the kind and amount of computer facilities available and accessible to the student researcher, and (3) the student's level of computer training.

Student researchers must be aware of the computer equipment available as they develop their research plan. This may influence to some degree the overall research design. However, a sound and viable research problem should not be altered merely to make use of, or perhaps avoid, using computer facilities. The soundness of the research problem should take precedence over the use of specific tools or techniques.

Conclusions and Prediction

The final stage in a research problem is to state conclusions based on the analysis of acquired data. All conclusions must be either carefully documented by utilizing pertinent specific data or, if based on the opinion or judgement of the researcher, they must be stated as such and the rationale, or logic, for these opinions clearly explained. It is important to realize that, if the research question cannot be answered completely, even though the research work was carried out in a sound and scientific manner, the results of the study are still valuable. In research work, it is not possible to answer the question or to come to positive conclusions in all cases. A research finding of no relationship among the hypothesized variables may add significantly to the body of geographic knowledge. It may save future researchers from wasting time by retracing the same steps of investigation, it may suggest more productive avenues of investigation, and it may dispel some common but erroneous notions held by others. In the final analysis, the success of a research problem is judged on the manner in which it was carried out rather than whether the conclusions were positive or negative.

If the research question has been answered, either fully or partially, the researcher can then project the findings and predict outcomes in situations similar to the research problem and matrix. A desired result of research is this ability to predict future situations. The researcher will find great satisfaction in his or her research work if the conclusions contain some element of prediction. Such prediction is also a purpose of science, although every scientific study does not have to be predictive in nature. It is further true that such prediction is usually limited to some time and place of the universe of the study. In the long range of science, universal laws are built from the bricks of limited predictions. Any time a norm, a law, or a history is established, a certain amount of prediction is possible for the universe studied. Their predictive value in a larger or otherwise different universe must be determined separately.

One of the best methods for the geographer to utilize for predicting is to construct a physical or conceptual model of an object or situation. This model becomes the basis for comparison, and deviations from it are apparent and measurable. For example, if the model of a city has a major business district near its center, any variation from this norm would attract attention. Geographers have devoted considerable time to constructing spatial models and adapting models from other sciences to meet spatial needs. An example of the latter case is the gravity, or interaction, model discussed previously. Obviously, the model is based on the law of physical gravitation. Such models, when properly modified to meet varying situations are quite helpful in spatial prediction.

The number and kinds of models, analogies, norms, and other predictive devices are so great that it would not be practical to catalog them in an introductory book on geographic research. Statistical methods and computer usage have become standards in some aspects of geographic research. In the next chapter, after introducing some elementary components of computer mapping and GIS, we demonstrate the use of GIS in geographic research.

Automation in Geographic Research: Searching Sources, Information Capture, Mapping, and Geographical Information Systems

.

Introduction

In previous chapters we presented various dimensions of scientific geographic research (problem definition, data acquisition, the spatial framework, data processing/analysis, data/cartographic portrayal, and conclusions). Computer automation has had an impact on geographic research methodology. This, of course, is true for most scientific inquiry. The major difference between geographic and other social science use of the computer lies in geographers' needs for data sets that link information to real-world locations. The advent of data processing, word processing, and computer graphics and image processing, particularly through computer mapping and geographical information systems, has revolutionized the way geographers are able to think about and solve geographic research problems. The purpose of this chapter is to identify first the general changes in executing geographic research given automation. Second, the basic elements of geographical information systems are discussed in terms of their links to research methodology. Finally, we provide a few examples of geographic research that utilize geographic information systems in real-world problem solving and theory building.

Automation: Changing the Ways We Pursue and Present Information

Throughout this text we have mentioned the importance of the electronic computer in performing one task or another that previously was accomplished by other means, including manual operations. It is useful to summarize several of these as they relate to capturing and storing, retrieving, analyzing, and portraying data for research purposes.

Library and Data Resources

The library is still the researcher's chief learning resource. If a researcher who died in the 1950s returned to life today, that person would recognize the familiar stacks of books and rows of journals in most libraries today. However, that researcher would find strange machines (computer terminals) and, in many libraries, the absence of conventional catalogs for retrieving books, documents, and data. Electronic searches for research materials are now common. Some libraries are networked to other libraries or agencies through computer. This permits access to off-site information at a cost less than purchasing and maintaining that information. One important example of computer-linked systems between libraries and off-site resources involves databases. Of particular importance to geographic research are those databases that have a geographic reference, or that report data for geographic units. For example, a researcher can query a large, off-site database for important demographic information for a specific state, county, city, or even some smaller geographic unit. The information typically is organized in report form,[1] including standardized information on population trends, age distributions, or age-income analyses, among others. Such data and reports, particularly between U.S. Census reports, provide information valuable to both the description of a proposed study area, and any changes within it since the last reported census, as well as data for comparative analyses desired between study areas. In fact, the same information and analyses are possible for microscale studies because these computerized databases use X,Y coordinates to produce data for any size area around a point in the United States. Thus, if a geographic researcher selects a site for a landfill and desires specific demographic data for one-mile and two-mile "rings" around that

---

1. The "MAX" system of National Planning Data Corporation, Ithaca, New York, is one such on-line system. There are numerous others that are utilized by libraries and by profit and non-profit agencies.

proposed site, a computerized system, often linked to a university library, can generate demographic data estimates and send them to the user over telephone lines in a matter of minutes.

We have also noted use of other technology in libraries. CD/ROM technology is one example. The entire 1990 U.S. Census is available in this format. Instead of bound reports on library shelves, a relatively small collection of CD/ROM disks, containing large volumes of information, not only require less storage space but are easily accessed by the researcher. This medium allows a researcher to create a digital file for computer analysis, rather than become involved in costly data entry that also may introduce errors into the new computer file. Examples of U.S. Census data previously published in bound volumes now available on CD/ROM format include the *U.S. Census of Housing and Population* and the *U.S. Census of Retail Trade*.

Another form of census publication involves the production of hard copy metropolitan maps. After 1990, all U.S. county and metropolitan maps were made available in digital format in a CD/ROM medium. We referred to these earlier in the text as TIGER maps. These permit the researcher to quickly access and manipulate maps, including those associated with research databases.

These are but a few examples of how the library and its resources have changed because of automation. Changes due to automation will most certainly continue.

Field Techniques

The library and its telecommunication linkages to other sources of data allow a substantial amount of geographical research to be completed using secondary or tertiary sources. However, geography is a discipline with a field tradition because many of its research problems are applied and require essential unavailable information. This is often the case when dealing with "microscale problems," dynamic regions, or "problems that require information that does not reflect itself as visible features (e.g., attitudes, shopping behavior, etc.) of the landscape."[2]

Automation has had a major impact on the methods of securing these and other data in the field. A few examples demonstrate this fact. Today, it is not uncommon for the researcher to carry a portable computer into the field for compiling field notes, for entering data directly into the machine, and for comparing computerized map features with corresponding real-world features. In all of these cases the researcher is attempting to save time and money, improving accuracy while in the field.

---

2. J. F. Lounsbury and F. T. Aldrich, *Introduction to Geographic Field Methods and Techniques* (Columbus, Ohio: Chalres E. Merrill Publishing Co., 1986), p. 86.

We also noted earlier that many types of measurements are made in the field, including temperature, precipitation, and distance, among many others. It has been common for field geographers to utilize instruments, such as radiometers (energy balance), theodolites (surveying), and plane tables and alidades (for field mapping). While these types of instruments are still used by some, others now use very sophisticated instrumentation for temperature, pressure, distance, and other measures. In such cases, computers play a central role and provide almost instant digital readouts of the variable measured. In certain cases, such as in surveying, lasers are now used to measure the distance between points. They provide the data in seconds.

Not only are field workers able to collect visible landscape features more quickly and more accurately today using automated methods, they also can utilize remote sensing technology to secure below surface data for microstudies. As Lounsbury and Aldrich note:

> "Ground penetration radar, proton magnetometers, and soils resistance measurement devices may be used to detect areas of soil compaction, near surface moisture relationships, subsurface cavities, archaeological sites, and urban data such as location of buried pipes and utilities. The sensed depth varies from one to several meters depending on site conditions."[3]

Automation contributes to more accurate field data, as well as providing remotely-sensed data that might otherwise be impossible to secure.

Statistical Analysis

We noted earlier that student researchers take for granted the time savings associated with word processing, which allows for immediate editing and routing of corrected pages to a printer. The same is certainly true for the availability of statistical analysis packages that permit thousands of manipulations in minutes, or even seconds. Historically, of course, all tabulations were accomplished "by hand." The electronic revolution made profound changes in statistical analysis for all of the sciences, beginning with the use of the mainframe computer. Later, programmable calculators permitted some calculations to be accomplished quickly in the lab or field, including measures of central tendency and some variance analysis procedures. However, today we can utilize even the more sophisticated statistical methods using software that operates on personal computers (PCs), and even laptops. Furthermore, some PC-based statistical software (e.g., SPSS) has a link to PC-based computer mapping software (e.g., SPSS-MAPINFO). Thus, one can analyze areas statistically and route the statistics to a mapping program that can plot the statistical scores. For example, one could calculate the mean

---

3. J. F. Lounsbury and F. T. Aldrich, p. 66.

income for all states of the United States, for all counties of the United States, and for all of its minor civil divisions and then map these average incomes as thematic maps at each scale (state, county, and minor civil division). Of course, more advanced statistics could also be employed and mapped at other scales.

Remotely-Sensed Images

Nowhere have automated methods had a greater impact on geography than in the portrayal of data from remotely-sensed images and in the production of computer-generated maps. Images and computer maps are part of a 1960s and 1970s computer graphics revolution that exploded in the 1980s. Early applications involved converting data into bar and pie charts and graphs. As we noted earlier, initial computer mapping software, such as SYMAP, operated on mainframe computers. These early maps involved the use of line printers that provided gray-tone patterns (not color).

By the mid-1970s new technology provided memory-intensive displays and lower-cost hardware with more memory. Color images and improved shading software followed. This was accompanied by the development of workstations. Soon three-dimensional surface mapping was possible using remotely-sensed data, such as radiometric data from aerial reconnaissance surveys. Maps could also be computer generated using digital data from LANDSAT and later from other satellites.

This new technology involves image processing, which provides an entire electronic snapshot of a phenomenon, or an entire landscape, through a process of "scanning," or transforming observations into computer code. The code is stored electronically and later is processed into an image. This technology is used to scan and store paper records such as conventional photographs, as well as images of landscapes. The American Express Company pioneered an application of this technology in customer billings in the 1980s. This involved images of their customers' charges (restaurants, hotels, etc.) and sending them to clients with their normal billings.

This same space-age technology has enabled the world to see "pictures" of other planets. Figure 7.1 shows the "Mountains of Venus," which result from data generated from the Magellan spacecraft flight in 1990. Such images of unseen landscapes not only permit us to see an area but also contribute to hypothesis generation for future research. In the case of Venus (Figure 7.1), mapping of the surface has changed a hypothesis about the age of its atmosphere. Because Magellan's "radar maps" of its surface showed no young, small craters, scientists now hypothesize "that the atmosphere has been thick enough long enough

FIGURE 7.1

(LAS—Jan. 25)—MOUNTAINS OF VENUS—NASA combined image and altitude data collected by the Magellan spacecraft to create this computer-processed, three-dimensional radar picture of mountains, cliffs and basins on Ishtar Terra, an Australia-sized highland on the planet Venus. The black strips are areas in which Magellan failed to collect picture information. The 2 ½-mile-high Lakshmi plateau is visible in the upper right part of the picture. (AP LaserColor) (h h61320ho/jol) 1991 (Planet is orange)
New York Times, Saturday, Jan. 26, 1991.

to prevent most small asteroids from reaching Venus's surface to gouge out. craters."[4]

Remotely-sensed data now play a vital role in geographic research, particularly when applied with additional analysis in geographical information systems. We will return to this thesis later in the chapter.

Computer-Generated Maps

Automated cartography employs computers to produce maps in various formats. A wide range of applications of computer-generated maps exists, including environmental, urban-economic, social, and engineering uses. Land use analysis, environmental impact assessment, and natural resource inventories are examples of environmental applications.

---

4. J. N. Wilford, "Images of Venus Reveal Volcanism—Craft Shows Role of Molten Rock in Creating Poisonous Gases," *The New York Times National Saturday,* January 26, 1991.

FIGURE 7.2

INFRASTRUCTURE INVENTORY

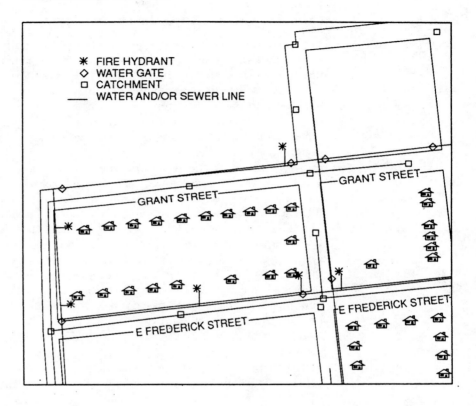

Engineering uses range widely and include inventory and analysis of infrastructure, as shown in Figure 7.2. Urban applications are numerous and include, for example, maps for housing inventories and analyses, land use maps for zoning, subdivision maps, and census maps for demonstrating an area's eligibility for or compliance with federal regulations. Numerous other examples are possible.

We now focus on an example of one set of computer-generated maps that demonstrate the advantages of computer-assisted mapping. Computer maps can be generated quickly and inexpensively at various scales. The digital cartographic files (illustrating the places or area boundaries) can be used with existing spatial data sets. Relatively inexpensive hardware and software assists us in producing the desired maps. Our example, census mapping using TIGER files, focuses on urban and socioeconomic mapping. The databases (cartographic and attribute) can be linked to more sophisticated systems that also perform environmental analysis, as demonstrated in the final section of this chapter.

Computer Maps and Census Geography

The generation of a census geography-based computer map requires coordinates (a set of X,Y cartesian coordinates that mathematically define the census areas) and a digital file that describes one or more attributes of the areas that are to be mapped. The U.S. Census Bureau manually produced and reproduced these maps for decades. Its maps of farm production, population characteristics, ethnic attributes, and other demographic distributions were well known and often used by geographic researchers.

A decision in the 1960s by the Census Bureau to develop geographic base files (GBF) for a mailout/mailback census provided one of the bases for what are now commonly termed TIGER ("Topologically Integrated Geographic Encoding and Reference" System) maps. The TIGER system updated the GBF/DIME (Dual Inventory Mapping and Encoding) system designed and developed in the 1970s. The GBF/DIME system was developed because it was apparent to Census planners that the old method of hiring enumerators to go door-to-door during the decennial census had become impractical. GBF/DIME became a digital computerized street map based on the entire metropolitan map series of the United States, and was supplemented by address ranges for every city blockface and by nodes that defined geographic coordinates for every intersection. This meant that a researcher could build automated maps from these files, including census tracts, block groups, or census blocks for any metropolitan area (see Figure 7.3a, b, and c). The researcher could also associate census data or local data to the local census maps. Thus, it was, and is, possible to examine distributions of any population attributes published by the U.S. Census (see Figure 7.4a and 7.4b), or to map local data at any level of census geography (Figure 7.4c). Figure 7.4c is a dot map of survey data from 1991. It depicts the locations of the two-family residential units in the city of Binghamton, New York in 1991. This local file contained the address of each unit and it was addressed-matched to its approximate X,Y location on the blockface, using the local geographic base file (GBF).

One of the advantages of the GBF/DIME system was its standardization nationally. For the 1990 Census, the Census Bureau updated the GBF/DIME files locally and incorporated them into the new TIGER system. This new digital geographic and cartographic system utilizes the data provided by the United State Geological Survey Topographic Maps (Digital Line Graphs) and the U.S. Census files in a single system. The result is the ability to produce census maps and data that are linked to

FIGURE 7.3(a)

1990 census tracts of the city of Binghamton, N.Y. There are 18 census tracts in Binghamton, N.Y. as designated by the U.S. census. Census data are published in paper and digital form for each census tract. Note the location of Census Tract 1 labelled as 100. Subsequent census maps in Figures 3(b) and 3(c) subdivide Tract 1.

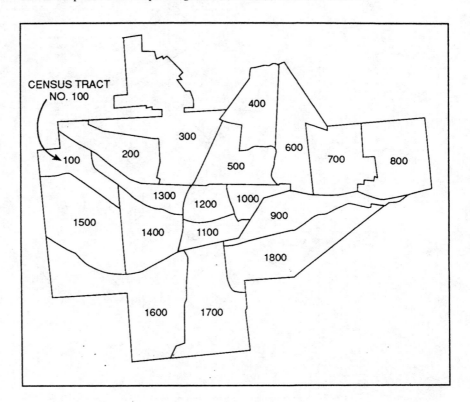

a useful digital database. Almost all commercial mapping software programs utilize TIGER maps for their system's basemaps and census boundary files. The result is relatively inexpensive computer files that permit low-cost computer-generated maps of any U.S. county, or smaller census geography areas. It is important to note that such computer mapfiles contain significant local features, e.g., railroads, streets, rivers, and political boundaries, including municipal and precincts, as well as census boundaries.

Geographical Information Systems and Geographic Research

Some researchers do not differentiate between computer-assisted cartography and geographic information systems. We chose to do so because

FIGURE 7.3(b)

Census block groups of Census Tract No. 1, Binghamton, N.Y. There are 4 block groups located within Census Tract No. 1. Detailed census data are published for each of those block groups. Note the location and shape of Block Group 1001. This block group will be subdivided into city blocks (census blocks) in Figure 7.3(c).

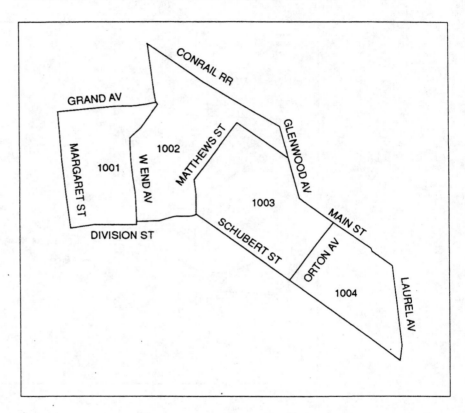

geographic information systems have greater analytical capability that often involves layering of data and statistical analysis. Before reviewing the basic elements of GIS, we differentiate between GIS and computer mapping systems.

Differentiating GIS and Computer Mapping: The Research Analysis Component

There are three applications of GIS and computer mapping systems in geographic research. Each system can be utilized in the first two applications, while only GIS is capable of performing the third. The first, and to date the most frequent application, is using a map to illustrate a set of spatial relationships. This can be as simple as illustrating the

FIGURE 7.3(c)

Census blocks of Census Tract No. 1, Block Group No. 1001, Binghamton, N.Y. Note that there are nine (9) census blocks in this Block Group No. 1001.

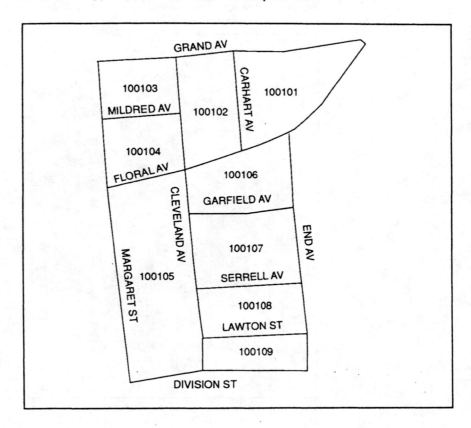

study area for a proposed research project, or portraying the distribution of utility lines in order to avoid them when digging for a repair (as in Figure 7.2). The latter use is sometimes termed Automated Mapping/ Facilities Management (AM/FM) applications. In either case, a database and map serves as a day-to-day filing system from which we can draw a map. A GIS or computer mapping system is capable of the portrayal function.

The second application generates a map to test a hypothesis. Frequently this involves the exploration of spatial data using a map to determine whether or not a spatial pattern exists. This requires tying spatial data to a coordinate file and generating a map. Again, either system is capable of generating a map using this process. The purpose, however, differentiates this approach from the first application.

FIGURE 7.4(a)

1990 Population Counts by Census Tract, Binghamton, N.Y. This is a thematic map of population distribution in Binghamton according to the 1990 Census of Housing Population. It permits a visual comparison of the census tracts populations.

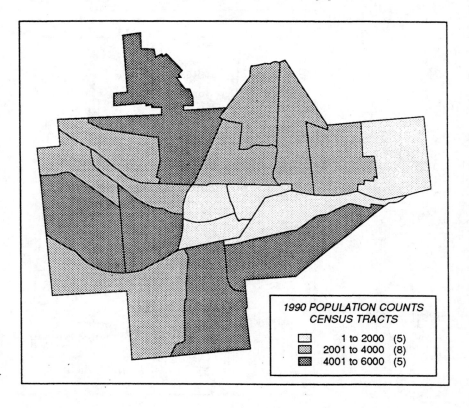

1990 POPULATION COUNTS
CENSUS TRACTS

	1 to 2000	(5)
	2001 to 4000	(8)
	4001 to 6000	(5)

Examples of this type of application could include an hypothesis that dilapidated housing is more highly concentrated near a city's center than near its periphery. Tying a housing condition (attribute file) to the census geography (coordinate file) of a community is required if the researcher desires an area map (or a dot map could be created using X,Y coordinates). Similarly, a researcher might propose locating a facility near a particular socioeconomic grouping, such as a senior center near an elderly population. In each case, the researcher mentally concludes that the hypothesis (or hypothesized location in the second example) is true or false. Notice that the computer software has not performed any analysis, rather it has created an illustration of a distribution by linking geographic and data attribute files. The analysis is performed mentally by the researcher when interpreting the computer-generated map.

FIGURE 7.4(b)

1990 Population by Census Block Group, Binghamton, N.Y. This thematic map provides a visual impression of the distribution of population in the city. While Figure 7.4(a) reported population for 18 areal units (tracts), this map illustrates population distribution for approximately 70 geographical units. Patterns within tracts are revealed.

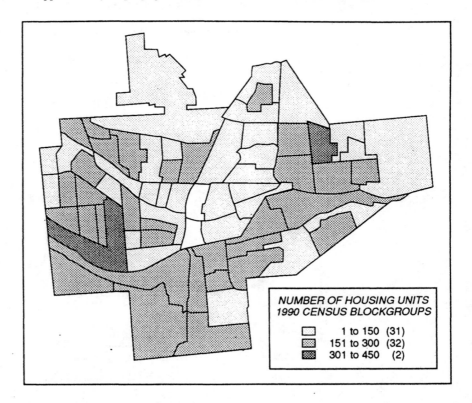

NUMBER OF HOUSING UNITS
1990 CENSUS BLOCKGROUPS

☐ 1 to 150   (31)
▨ 151 to 300 (32)
▰ 301 to 450   (2)

The final application involves using a system to test a hypothesis through the spatial analysis of data. Often such applications give rise to new hypotheses. Before discussing the analytical nature of a GIS that makes this application possible, we review the standard elements of any GIS.

GIS Elements

The software and documentation of any system to be termed a GIS must meet minimum requirements. A GIS must be able to capture and store, retrieve, analyze, and display data in a spatial format (locational reference is required).

FIGURE 7.4(c)

The Distribution of Two-family Units, Binghamton, N.Y., 1990. Note that this map illustrates the location of print data (2-family structures) by census block within Census Tract No. 1, Block Group 1001. Thus, the reader can note that Block No. 100103 has no 2-family units, while Block No. 100102 has five (5) 2-family units.
Source: City of Binghamton, N.Y. Assessor's Data File

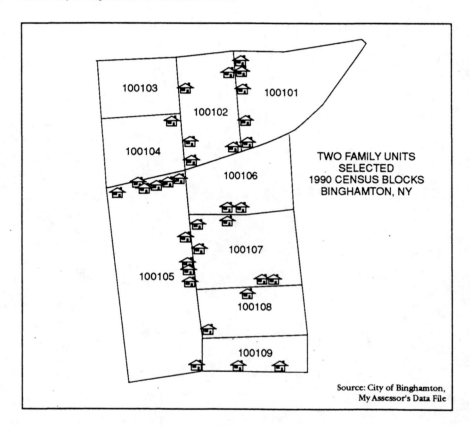

Data Capture and Data Storage

Typically, we refer to input devices when speaking of the means used to input data into GIS or computer mapping systems. These input devices can be a computer terminal or workstation. Data can be input from a computer keyboard or entered via a digitizer, which is a device that creates a digital map from a conventional paper map. We also noted that image processing provides digital data for use in a GIS. In short, any map-producing system needs a coordinate file or scanned map image and a data file representing attributes of the map. Whether census data or soils data, each data source must be entered into the system. Some might be read in from an existing computer tape or disk, other must be created (or captured) for the first time.

All data are prepared or transformed into a computer-compatible format and stored in a particular way (whatever form is dictated by the hardware and software). In a GIS, it is common to store data by layers, such that there are several environmental layers (soil, vegetation, water, etc.), as well as other layers (e.g., highways, demographics, boundary files, etc.) (see Figure 7.5). Computer-mapping systems may or may not permit data layering.

Data Retrieval/Display/Portrayal

Data must be retrievable from storage and must be capable of being displayed on a computer screen, or as hard·copy maps and graphics. The spatial data can be portrayed as point, line, or area data. In addition, some systems portray surfaces, i.e., display of a vertical property over an area. An example is elevation illustrated over a study area. As noted earlier, this requires an ability to tie specific attribute data to locations.

While electrical impulses create almost instant maps on a video screen, it is desirable, if not necessary, to create hard copy output of maps, graphs, charts and other computer graphics. Therefore, a GIS or computer-mapping system must be capable of routing its electronic images to equipment that draws maps (we call these output devices). Output devices include printers, such as inkjet or laser printers, and plotters, both pen and electrostatic, that result in a wide range of quality in the production of maps in both color and black and white. There are other output devices, such as copiers and photographic film writers, but it is not our purpose to explain such devices.

Spatial Analysis Capabilities

We maintained that analysis capability differentiates a GIS from a computer-mapping software package. Up to this point in our discussion, both types of software could function to capture, store, retrieve, and display data, although admittedly perhaps with very different levels of quality and accuracy. A computer-mapping package, however, has very little or no analytical capability. It must rely on supplemental database management and statistical packages to combine and otherwise manipulate data elements and to statistically analyze spatial data. A true GIS is a "system" that includes not only map files and other spatial data in digital form, and attribute data stored in layers, but also incorporates analytical software. Campbell notes the "usual capabilities" of any GIS as:

1. Comparing the distribution of two types of data within the study area;
2. Searching for a selected set of characteristics that occur together, either with or without some other specific characteristic;

FIGURE 7.5

GIS layers.   Credit: Dennis Anthony, SUNY Binghamton, Department of Geography, Computer Mapping Laboratory.

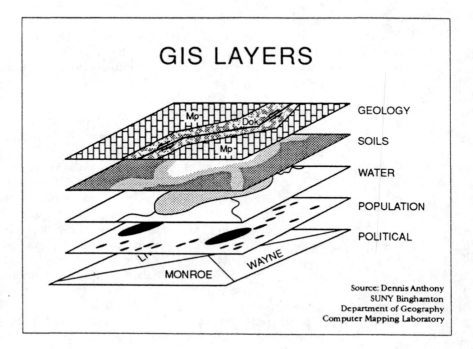

GIS LAYERS

GEOLOGY

SOILS

WATER

POPULATION

POLITICAL

MONROE   WAYNE

Source: Dennis Anthony
SUNY Binghamton
Department of Geography
Computer Mapping Laboratory

3. Searching for the nearest neighbor of a specific feature, with added criteria regarding characteristics;

4. Handling comparisons between data recorded at different scales or on different projections.

The basic operations that support these capabilities include reclassification, overlaying, distance functions, and modeling.[5]

Manipulations of data are necessary when performing any quantitative analysis. It may be necessary, for example, to reclassify individual ages to a classification of young, middle age, or elderly. Similarly, it may be desirable to change a multilevel soil classification to a more manageable one with fewer classes.

Additionally, it is necessary to perform numerous calculations when conducting a statistical analysis. (This should be clear from the data analysis chapter of this text.) Modelling data, as we noted in Chapter 6, involves describing a set of phenomena as hypothesized in the interre-

---

5. J. Campbell, *Map Analysis: An Introduction* (Dubuque, Iowa: W. C. Brown Publishers, 1990), pp. 267–268.

lationship of a selected set of variables. GIS modelling uses the GIS database to create information about possible outcomes based on particular conditions. We discussed the gravity model, which defines spatial interaction in Chapter 6. You will remember that the interaction between markets is predicted as equal to the product of the size of the two stores (or cities, depending on the type of application) divided by the distances between them. This retail gravity model, built into some GIS algorithms, requires the measurement of distances between all points under consideration (e.g., stores, cities) and a set of mathematical calculations ($i = \dfrac{P_1 P_2}{d}$ for 2 stores). Some GISs also graphically display market share and client intensity as market rings around each store. This type of data manipulation for analysis distinguishes a GIS from less sophisticated computer-mapping software.

To return, then, to our final application, it is the use of GIS for hypothesis testing, often times using modelling, that sets GIS capabilities apart from computer-mapping software. In the retail gravity model example above, the statistical model tested the hypothesis and the map was used to illustrate the results. Both are accomplished by a true GIS. A GIS can be used to formulate and test hypotheses.

Hypothesis Testing: A Theoretical Analysis Using GIS

Given a GIS with multiple layers of environmental and socioeconomic information and the ability to utilize image data, a great range of applications for hypothesis testing is available to geographic researchers.

GIS construction is driven by the users needs and goals. The more complex and demanding these are, the more complex and expensive the GIS system. A GIS designed for multiple purposes for a nation or state will be far more complex than that of a small city. The data needs alone will be quite different, as will concomittant storage space, maintenance, and other needs. In short, any GIS design must consider hardware and software options, as well as the following specific issues: data needs and sources, geocoding procedures, the number of separate layers to be created, and the analytical functions required to analyze the data, which implies a plan of action, or strategies, for use of the GIS.

As we noted earlier in this text, research problems dictate data needs. Let us suppose for our theoretical case that the range of research questions that will drive our GIS relate to environmental variables, specifically erosion, deposition, and water quality issues. The existing body of knowledge suggests that at a minimum we need to secure data for our study area that includes soil, vegetation, surface and ground water, slope, and land use (implying human impact). We must then determine where to get these data. Sources could include the USGS topographical maps, data from the Department of Environmental

Conservation, the Environmental Protection Agency, and maps from the Soil Conservation Service. We also might search for available aerial photography and remote sensing images for land-use data. Some will even exist in digital form. We might also have to create data from samples taken in the field, particularly if the water quality data are unavailable or are outdated. We will enter these data into the GIS.

In constructing the GIS, one of our first decisions is the geocoding decision. We must decide which map coordinate system to use. Our choice will determine how easy it is to use and how precise and accurately located our data will be. No single system offers all desired benefits; there are trade-offs. The UTM (Universal Transverse Mercator) system, for example, uses 16 digits to store a location to a 1 meter precision. The system can also be modified such that, when a resolution of 10 rather than 1 meter is needed, a digit can be eliminated. It also has the advantage of permitting geometric computations on spatial data that result in minimal errors for distance measures over regions. It has the advantage of being universal. One disadvantage of this system is that, compared to other systems, it is less accurate for small areas.

Another coordinate system used for geocoding geographic data is the State Plane Coordinate System, which typically is based on individual state maps in the United States. The result is that the number of zones in each state varies by the size of the State and the projection used. The State Plane System is based on feet (not meters) and thus is more accurate than the UTM System for small areas. It suffers the disadvantage of lack of universality. Other systems are also available and have advantages and disadvantages that must be weighed by the designer before selecting the geocoding system.

Another choice to be made is whether the data to be displayed by the GIS should be captured in a raster or vector format. Line drawing displays are called vector systems, while a raster display involves portraying data as cells based on small squares, rectangles, or dots, referred to as pixels. The major advantage of the vector system is the ability to accurately draw straight lines. Thus highways, rivers and other lines are displayed accurately and in a visually pleasing way. Raster displays project only those pixels at grid points. The result may be jagged lines. One disadvantage of the vector system is that because only a limited number of lines can be drawn, map complexity is limited. Another big difference between the systems lies in the time required to digitize. Scanner systems quickly scan an image and record binary values for every pixel. Remotely-sensed images, digital photographs and digital map files are produced quickly and relatively inexpensively. Vector systems digitize maps by recording X,Y coordinates for vertices of every line on the map. Thus, the time required to create a highway map is substantial, particularly because highways frequently change direction, requiring numerous vertices.

The purpose of the GIS will dictate which system is necessary. In fact, some commercial GIS systems incorporate both systems. In our theoretical GIS, because we will want to look at the combined effects of slope, vegetation, land use, etc., on erosional patterns, and because we will use remote sensing data, we will need a raster-based system.

We will also need separate layers for each of these variables (slope, vegetation, water, land use, etc.) so that we can examine various combinations of variables at any given time. We are using a small study area in rural Pennsylvania. These facts result in our decision to use a state plane coordinate system with raster digitizing.

In addition to layering, this GIS must have data analysis capabilities so that we can perform a number of functions. For example, we will want to model actual and expected distributions of erosion patterns, as well as to perform some standard statistical measures, such as a t-test.

In summary, we will design a number of research projects dealing with surface erosion and water quality. One such research problem will involve testing hypotheses regarding particular patterns of erosion (expected). We have noticed that erosion rates vary over a large area and are concerned with the impacts of slope, soil, and vegetation regimes on these erosion rates. We identify several types of slope, soil, and vegetation regimes in the area. The research proposal discusses the data and its sources. The methodology includes putting the data into a raster-based GIS so that we can analyze the combined effects of soil, slope, and vegetation on erosion rates. It also includes the necessary modelling procedures and statistical tests to be applied to test the theoretical erosion patterns. In this way, GIS play its proper role—it is the geographic researcher's tool to test a hypothesis for theory development. It provides the data, analysis, and portrayal functions necessary to complete the study.

As noted in Chapter 1, Geography has both theoretical and applied pursuits. We now briefly examine two research applications that demonstrate GIS ability to combine and analyze different data sources. A single automated system executes all data needs: capture, storage, analysis, retrieval, and portrayal. The first example illustrates GIS utility in an applied context for resource managers. The second demonstrates how GIS technology is incorporated into a research design for the purpose of theory development.

GIS in Applied Geography

GIS applications are occurring around the world. Most of these applications are in resource management and environmental planning. In the United States, for example, applications have become common in forestry, fish and wildlife, energy, and environmental planning at the local, state, and federal levels. More recently, GIS has been applied in urban economic and social planning. The Department of Housing and Urban Development, for example, has completed demonstration projects related to housing quality, program eligibility and compliance patterns, and housing strategies using GIS. The private sector also employs GIS for business planning and target marketing purposes. Since most applications have been environmentally-oriented, however, we chose the following example for a GIS application.

Resource managers and planners are charged with locating and managing existing and potential problem areas. Typically, they are interested in securing data from remotely-sensed images, UTM projections, black and white photography, existing maps and field data to identify and analyze these problem regions. As in our theoretical case, coverages typically include slope, land use, land forms, soil attributes and types, water and drainage, highways, as well as human attributes such as population density.

In many developing countries it is critical to manage important natural resources. Forest reserves in Botswana are one example. In this specific context, Nellis, et al., noted:

> "To adequately manage these critical resources will require more accurate information on the extent of elephant populations and the associated impact of elephants on the vegetation. Space Shuttle Photography offers one source of spatial data, that when combined with other geographic data, can be of significant utility to natural resource managers."[6]

Using a GIS and layers of data secured from a number of databases, these researchers tested the hypothesis that the greatest elephant-induced forest damage was adjacent to the Chobe River in Botswana. The analytical capabilities of their GIS included the use of video-enhanced space shuttle photography (see Figure 7.6a), which converted a visual photo to a digital photo. It also applied two statistical techniques, clustering and classification, to enhance the differences among landscape areas of the photo. The result was six distinct landscape classes (see Figure 7.6b) that could be illustrated using gray tones. They then compared this classification scheme with layers of environmental data for the study area, including soil type and topographic position. They also used other statistical measures for each classification (zone) to illustrate elephant-induced impact on the vegetation. Thus, these researchers could describe environmental zones and elephant impact zones around the Chobe River.

> ". . . (it was) determined that the vegetation units represent a topo-sequence from bare soil (and heavy elephant-impact areas) near the Chobe River (lighter tone), to more dense teak forests at slightly higher

---

6. M. D. Nellis, et al., "Planning as Applied Geography: Specialization and Integration," *Papers and Proceedings of Applied Geography Conferences*, Vol. 13, 1990, p. 14.

FIGURE 7.6(a)

Space Shuttle photograph of northern Botswana and a portion of the Caprivi Strip. The northern border of this portion of Botswana is defined by the Chobe River (A), which merges with the Sambezi River. The Kasane Forest Reserve (B), as part of the Chobe District is part of the Sambezi Teak Forest Association. Credit: M. D. Nelbis, et al., and Applied Geography Conferences. Published in Volume 13, *Papers and Proceeding of Applied Geography* Conferences, Vol. 13, 1990, pp. 12 and 14.

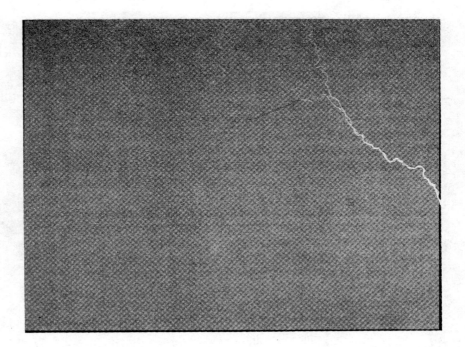

elevations, and greater distances from the river (darker tone). The spatial banding zonation away from the elephant's water source is very evident . . . although elephant impact on vegetation is significant, the pattern of impact is quite variable within a localized area. . . ."[7]

This is a good example of a GIS designed to address specific environment-oriented issues. It utilized a raster system to identify and analyze combinations of variables in relatively small areas. The GIS had the capability to integrate the typology of different databases for the specific study area. It also has strong layering and analytical capabilities needed to solve this applied research problem. The GIS tested a hypothesis with a client in mind, the resource managers of Botswana. The results provided useful insights into areas of varied impacts.

7. M. D. Nellis, et al., p. 13.

FIGURE 7.6(b)

An unsupervised cluster routine applied to a video digitized Space Shuttle photograph of the Chode District, Botswana. The gradations in gray tone relate to the major landscape units within the Chobe National Park and surrounding area. The lightest tones at A represent the area adjacent to the Chobe River and most heavily impacted by the elephant population. Darker tones represent higher densities of forest cover.

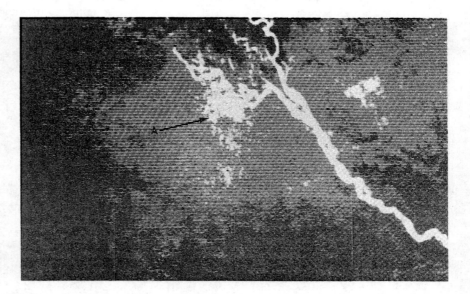

GIS and Theory Building

GIS and other tools assist geographic researchers in their pursuit of theoretical truth. Often in science we are unsure of the roles causal factors play in some process. Complex problems require carefully considered, complex explanations. As a result it is common for analyses to incorporate multiple data sets and a wide range of methods to test theoretical propositions. More and more GISs are contributing to such complex methodologies. Such was the case in the research by Dobson, Rush, and Peplies, which sought to clarify the relationship between landscape processes and lake acidification (Dobson, et al., 1990). In the author's words:

> "Our approach combines GIS and digital remote sensing with traditional field methods. The methods of analysis consist of direct observation, interpretation of satellite imagery and aerial photographs, and statistical comparison of two geographical distributions—one representing forest blowdown and another representing lake chemistry."[8]

8. J. E. Dobson, R. M. Rush, and R. W. Pepliles, "Forest Blowdown and Lake Acidification," *Annals of the Association of American Geographics*, Vol. 80, September, 1990, pp. 343–49.

The process of lake acidification is quite complex. Many factors could contribute to the process, "including atmospheric contributions such as acid deposition, terrestrial features such as geology, aquatic variables such as the dwell time of lake waters".[9] Of particular concern to these researchers is the geography of lake acidification and the variation of acidification severity. They specifically focused on the long-term consequences of a devastating storm, termed "The Big Blow," on the Adirondack Region. The Big Blow was a storm of enormous proportions, blowing down "25–100 percent of the trees on 171,000 acres within the Adirondack Park".[10] Blowdown theory suggests that an event of this magnitude changes lake water chemistry by massively increasing the volume of dead biomass in the area and by changing local hydrologic flow.

Dobson, et al., sought to determine accurate estimates of blowdown and to test hypotheses regarding the relationships between blowdown intensity and the lake chemistry within selected watersheds. Among their specific hypotheses were the following for 208 Adirondack Headwater lakes:

H_{01}: There is no interrelationship between forest blowdown and lake Ph (a measure for acidity/alkalinity) reading.

H_{02}: There is no difference between category of blowdown damage and Ph reading.

H_{03}: There is a zero correlation (no relationship) between Ph reading and the extent of storm damage.

In each of these null hypotheses the authors posit that acidity is independent of some measure of blowdown. The alternative hypotheses suggests that blowdown (its existence or magnitude) is related to lake acidity as measured by a Ph reading.

GIS was utilized in order to test these hypotheses. The three separate statistical analyses required multiple data sets, which were generated from different sources. The researchers used New York DEC maps of blowdown occurrence (created in 1950) that were based on air photography. They also utilized USGS topographic maps (1988) of the Adirondack region. Standard GIS procedures were used to digitize overlays of blowdown and watersheds. The former polygons were taken from the DEC maps, the later from the USGS topographic maps. Polygon intersections were calculated using an Oak Ridge National Laboratory GIS. Data were tied to intersecting polygons. These illustrated blowdown occurrence and intensity, the percentage of watershed area damaged by The Big Blow. Other data were also incorporated into the

---

9. IBID, p. 343.
10. IBID, p. 347.

GIS, thus creating additional layers of information for the study area. Various combinations of data could be identified for locations within the study area by the GIS. Statistical analysis could be applied to these data. For the first hypothesis, Dobson, et al., applied a chi square test to evaluate independence of lake acidity and blowdown. The second hypothesis was tested using analysis of variance, which statistically test whether or not two groups (categories) vary similarly in terms of another variable, in this case whether or not categories of storm damage (e.g., high damage and low damage) vary similarly in terms of their Ph readings. The final hypothesis discussed here used Spearman Rank Correlation (see Chapter 6) to test for a zero correlation (no correlation) between a lake's Ph reading and the extent of storm damage in the watershed.

In all three cases the null hypothesis had to be rejected. Thus, lake acidity is interrelated with the storm damage of The Big Blow. Notice that the GIS permitted these analyses because it was able to combine data elements for a particular location after the data were captured from various sources. This ability makes GIS one of the geographic researcher's most useful tools.

Summary

Automation has changed the way science is conducted. Like the other sciences, Geography has benefitted from the electronics revolution. Library resources are identifiable almost immediately and due to remote access we have access to even more documents and data than ever before. Computers allow us to see information more quickly and to be transferred from one medium to another, which makes analysis faster and often cheaper.

Field methods have also changed due to automation. We now can take a portable PC into the field to collect data. Also, new sophisticated electronics take faster and more accurate recordings in the field. An example is surveying. Remotely-sensed images are another example of data generated by advanced technology that permits researchers to see formerly "invisible" landscapes.

Statistical analyses are also more quickly performed using electronics. The same is true for electronic graphing and computer mapping. GIS, however, is one of the newest and most promising technologies to influence scientific research in decades. Its ability to capture, store, retrieve, portray, and analyze information is unique. Applications of the technology to a wide array of problems has demonstrated its utility. Real-world problems in planning, resource management, engineering, and marketing are a few of a many to which GIS has provided solutions. Its greatest strength lies in its ability to combine attributes for any specified location for the purpose of hypothesis testing.

Writing Geographic Research Reports

Research that is unrecorded, or inadequately recorded, has little value to anyone but the researcher. Undoubtedly, much research has not been made available to others because the study has not been written, or because it has been written in an unsuitable form. Writing a geographic research report is often difficult for the beginning research student, but it is an important part of the research project. It must be approached with the same sense of scholarship as the researcher feels toward the other components of the research problem. It requires skill, organization, insight, and hard labor. A part of this task is the proper arrangement of the physical materials needed in order that the mind may be free to concentrate on the subject matter.

The Work Area

The fact that writing involves a considerable amount of creativity is all the more reason that the tools and necessary equipment should be assembled in advance and be readily accessible. This necessitates having all tools and supplies in an established location. It is not as important that there be a particular arrangement as it is that the equipment always be in the same location. However, as the student geographer should know by this time, organization of space is important at the microscale as well as at the macroscale.

The writing of the geographic research paper requires certain basic supplies. Some of these may be used up rapidly, and so they should be available in sufficient amounts to ensure that there be no need for leaving the work desk to obtain more. These supplies include pencils, pens, a ruler with both metric and English scale, and sufficient paper. It is advisable to have some extra note cards along with the supplies.

It is likely that the writer will need extra base maps and colored pencils for making preliminary maps. Frequently, ideas occur during the writing process, and it behooves the writer to record his thoughts immediately. Existing maps should be included among the working supplies. There should also be available source books, such as a good dictionary, a word guide for spelling, a thesaurus for word choice, a grammar text for form and structure, atlases pertinent to the subject, a cartographic manual, special writing or style manuals determined by the school or by the ultimate authority who will judge the work, and basic technique materials, such as qualitative tables. In short, any reference book or source material to which the researcher may refer frequently should be readily accessible.

Preparing to Write

With notes sorted and supplies and sources gathered, the student is ready to go into seclusion preparatory to the actual writing. Scholarly writing does not come easily to most people. One of the rules for the writer to remember is that his first task is to communicate with the reader. No matter what else he has done or will do, it will be lost if he does not honestly communicate. This includes the use of clear, simple sentences, and the avoidance of jargon. He should always remain objective, write in the third person, and avoid the editorial "we." This does not imply stilted or awkward usage, whether writing in the active or in the passive voice. Since the study has already been conducted, the past tense is mainly used throughout the report; but the present and future tenses may be used where appropriate.

It is not by accident that the phrase "a scholar and a gentleman" has been so widely circulated. Today, of course, the term "gentleperson" is more appropriate. The scholar is helpful, courteous, and considerate. If an error is found in the work of another, it is not introduced into the research report unless it has a bearing on the problem. If it is a part of the study, the original source and the accurate correction should be given. Modesty is a virtue highly recommended, especially for beginning scholars. Albaugh states that an overly positive, dictatorial air, inappropriate in any scholarly writing is unforgivable in a graduate thesis.[1] The same would be true of research reports in general. Scholars do have opinions and state them; but while they are positive, they should use due caution. If a conclusion appears to be justified from the evidence, they would probably say "it appears that" rather than "it is obvious." When the opinion is not their own, scholars are careful to give credit to the source.

The First Draft

The writing of the first draft of a geographic research report can be an extremely pleasant period for the researcher. He is at the stage where, like the artist at his easel, he sees the planning and work of his project take visual form. The pleasure he derives from his own creation can be a great reward for any scholar. However, the writer should prepare to give the composition his full attention because interruptions at this point may be destructive to his creative capacity.

The first or rough draft is intended to be the framework within which the researcher organizes and records his findings and ideas in a systematic form. It is the rough sketch of the final creation, and is not intended to be widely circulated. Consequently, it should be written as rapidly as is consistent with good thought. It is hoped that style and usage are innate capabilities with the writer since he should not devote extensive time to such matters while composing the first draft. If mistakes in grammar or style occur, they may be noted; but detailed editing should not be done at this time. All writing should be double- or triple-spaced in order that inevitable corrections, additions, or modifications may be made later. It is desirable that only one paragraph be written to a page so that the order of thoughts may be rearranged later. Quotation note cards and other notes may be stapled directly to the appropriate manuscript page for convenience. Footnotes are essential, but a short form should be used because little time should be taken in making such notations. The name, year, and page should certainly suffice here.

Exactly how the written draft is handled varies with the length and type of geographic research report. A term paper may be submitted to the instructor for tentative approval, and a graduate thesis may be submitted one chapter at a time. In any case, the first draft should be set aside to "cool" for a period of time. After an interval, it is ready for critical review; and the task of the second writing begins.

---

1. Ralph M. Albaugh, *Thesis Writing* (Totowa, N.J.: Littlefield, Adams & Co., 1962), p. 11.

The Second Draft

The term "second draft" is really a euphemism for a considerable amount of writing that is necessary between the first and final draft of the geographic report. While it is common to think of writing three drafts of the research paper, one should realize that many more may be necessary. Seasoned writers frequently rewrite a page ten or more times before it is satisfactory. Certainly, the beginning writer should not be disappointed if his second writing is not satisfactory in all respects. The second draft is a refining process that advances the manuscript from the rough recording of facts and ideas to the stage where the various components of the report mesh and can be viewed in perspective.

It is here that the writer works on cohesion of sentences and paragraphs. Reorganization often is necessary so that the sequence of thoughts is presented in the most logical manner. Errors of detail which were allowed in the rough draft are corrected, and unclear points may need elaboration and clarification. Verbosity and clichés of thought or expression are pared from the manuscript.

In a geographic report, the rough draft includes illustrative materials (sketched hastily or made with pencils) which need correcting and revision. These include maps, charts, and diagrams used or referred to in the first draft. Like the painter who is filling in the color and correcting the format from the preliminary sketch, the research writer now sees the creation taking the form that gives pleasure to the scientist—a research report that communicates from writer to reader with simplicity and power.

The Final Draft

The final draft would be anticlimactic were it not the culmination of the entire project, the end toward which the research project was aimed. Even so, it is primarily the final editing of the completed report, the refining, perfecting, and clearing of miscellany.

In the final drafting, an attempt is made to obtain perfect sentence structure and spelling. Quotations and footnotes are given their last scrutiny and approval. The final bibliography is assembled, and all maps are reproduced for inclusion in the completed report. Since this is the final form of the report, page numbering, indexing, and making content tables, and all similar items must be completed at this time. When the final draft is completed, the manuscript is ready to go to the typist or typesetter, as the case may be, to be put into its final form.

Types of Geographic Reports

The general content of the scientific research report is essentially the same regardless of its magnitude or purpose. When the research steps described in previous chapters have been carried out, the purpose of the written report is to record accurately the findings and results of the project. The scope and length of a given report will vary, depending upon the research problem and its *raison d'être*. In most cases, a research report will fall into one of seven broad groups: (1) academic theses and dissertations, (2) articles in professional journals, (3) articles in semi-professional magazines, (4) technical reports, (5) textbooks or chapters in texts, (6) term or seminar papers, and (7) proposals requesting financial or other forms of support or approval. The first six types of reports are concerned with research that has previously been completed. The latter type describes a proposed research project. However, a sound and detailed proposal will require a great deal of investigation and study, and in this sense, it is a type of research report. Each of these types of research reports are written primarily for a specific audience and, therefore, will vary in some respects in content, style, and organization. Theses and dissertations are the only types of research reports that always include all the components of a formal and rigidly organized report. For this reason, this type of report merits special consideration here. Further, the master's thesis is the most universal requirement in American colleges and universities today, and as such, will be treated here as a special case and model.

Theses and Dissertations

In most cases, the major difference between a master's thesis and a doctoral dissertation is one of magnitude. The dissertation normally is concerned with a more sophisticated problem, takes longer to complete, and is subject to the highest standards of research and scholarship. Normally, it deals with primary sources and the acquisition of new data to a greater degree than a thesis, as well as to new methodologies and models.

For the average graduate student, the master's thesis is an intense and meaningful research effort that culminates academic progress toward the master of arts or master of science degree. Through this scholarly work, students not only succeed in demonstrating their own mastery of research tools and methods, but they join the intellectual ranks of scholars reaching back to the medieval universities of Europe. To be sure, the idea of the thesis has changed since its beginning in those early universities. At that time, the thesis was a proposition supported by the fledgling scholar. Academic ability was demonstrated by public defense of the proposition, essentially a display of speaking and debating ability, although knowledge of the subject was also necessary.

This system prevailed in European and American universities until fairly recent times when emphasis was shifted to the research aspect. Research is the dominating theme of the thesis today.

The master's thesis is commonly regarded as an intensive research work that allows the writer to demonstrate the ability to conceive a problem and proceed to its solution. It embodies both the processes and the results of investigation. It is usually different from any other research that the student may have previously undertaken, although the difference may be more a matter of degree than of kind. For most, it is the first attempt to produce a scholarly work that is published, bound, and shelved. Thus, it bears the writer's name, and for all time, the quality of his scholarship will be linked to that work. In the event that he or she does not publish subsequent studies, the thesis will remain the major measure of the quality of the scholar.

In some respects, the thesis resembles term papers required in undergraduate classes. It is an individual work that is systematically pursued and reported. The thesis is usually longer than the term paper because it represents a research effort extended over a greater period of time. However, the length is not necessarily a measure of the quality of the thesis. In fact, the thesis problem may be more limited and specific than the term paper subject. The term paper often has a general topic and is actually no more than a guiding statement or "theme" around which the research focuses. The thesis has a specific problem for which a solution is sought. The emphasis of the term paper is commonly placed on demonstrating the skills of adequate library research and its proper reporting through a written paper. The thesis relies more heavily on the use of primary sources both in and out of the library. Both must be reported in a scholarly manner, but the thesis must do much more. It must show the reader the processes and reasoning, as well as the results of the research, and it must evaluate and present the evidence which leads to the conclusion.

In summary, the master's thesis today in American education is a major criterion in measuring the student's scholarly abilities. It is a tangible example of scholarship. With it as evidence the student stands ready to be judged by the masters in his or her selected field of study, masters who will determine if he or she qualifies to join their ranks. Through the thesis, the student demonstrates an ability to conceive and state a problem in geography and can proceed to locate and analyze the data that are related to the problem. Mastery of the analytic tools and techniques necessary for scholarly success in geography should be adequately evident. Conclusions, properly drawn from the thesis study, should be stated in a manner that demonstrates a high degree of reasoning ability and command of the scientific method of inquiry. It is worth repeating that it is the research method, especially the reasoning displayed in its application, which is of major concern to those who judge a petition to become a master scholar.

Organization of the Thesis or Dissertation

The thesis or dissertation, when it is complete, is presented in the form of a book. Although the following discussion is concerned with the master's thesis, it is equally appropriate to the doctoral dissertation. The organization is presented in three sections: (1) the preliminaries, (2) the text, and (3) the reference material. Each section is vital to the whole, and the finished work is not complete without proper attention to all its parts. An abbreviated suggested outline of the proper thesis organization follows. Some geography departments might vary the order slightly, but such variations are usually minor.

I. The Preliminaries
   A. Title
   B. Submission
   C. Acknowledgments and/or preface
   D. Table of contents
   E. List of tables
   F. List of illustrations
II. The Text
   A. Introduction
   B. Substantive report
   C. Conclusion
III. The Reference Materials
   A. Bibliography
   B. Appendixes

If the writer keeps the primary objective of the research work clearly in mind, the organization of the material becomes quite logical in its arrangement. Following the standard preliminaries, the first part of the text introduces the problem and establishes the ground rules to be followed in its solution. In the second and third parts, the data collected are analyzed, synthesized, and evaluated in order to test the hypotheses as solutions to the problem. The final part of the report is a listing of sources referred to in the study, as well as additional material considered pertinent and useful to future researchers.

The Preliminaries   The document begins with the title page, just as a book begins with the title. Of course, the bound copy has a cover and perhaps one or more blank pages before the title; but the first page of the thesis, counted as page one in small Roman numerals, is the title page. The exact content and form of this page varies slightly, depending upon the individual institution. Commonly, it contains not only the title and the author's name, but also the date the degree will be awarded, the school, and the degree for which the thesis is submitted as a partial requirement.

The title page is followed by a page called a "submission statement." This page contains the title and the author's name again. The date of the submission page is the date of the thesis approval by the faculty; usually, the date of the oral defense is the one designated to be shown here. Following the date, space is left for the thesis committee members and, perhaps, the dean of the graduate college to sign approval. This page is part of the finished thesis and states that the faculty has approved of the scholarship demonstrated in the written report. It may thus logically be assumed that the thesis represents their standard of excellence for a master scholar.

In many books, a statement of the aims and nature of the work to follow is made in an opening statement called the preface. Since the thesis includes a careful appraisal of the purpose, scope, and anticipated results of the research in an introductory chapter, it is not unusual to find this material in a preface. In its place is usually found an acknowledgment page which is a short, simple, and tactful statement acknowledging those persons who contributed significantly and beyond normal expectations to the completed work.

The preliminary section of the thesis is completed with the table of contents, a list of tables, and a list of illustrations. The pages allow the reader immediately to observe the organization of the work and to locate the pages that contain the parts of the study in which he might be interested. Since the thesis does not commonly contain an index, the reader must rely heavily on the page number references in the table of contents and the lists of illustrations and tables.

The Text The heart and purpose of the thesis is the text. It is essentially composed of an introduction, the substantive report, and a summary or concluding chapter. There is no set number of chapters for the three parts of the text. It is usual for the introduction to be presented in one or two chapters. The concluding section is usually one chapter in length. The substantive report itself may be one or more chapters, depending upon the logical division of the material into identifiable divisions. The parts of the thesis text, presented in a desirable and acceptable order, are shown below. However, no set order of presentation is required, and practice varies widely.

A. Introduction
   1. Justification and need for the study
   2. Precise statement of the research problem and research area
   3. Precise statement of the hypotheses
   4. Definition of terms
   5. Limitations of the study

6. Review of the pertinent literature
7. Organization of the study
B. Substantive Report
    1. Presentation of data
    2. Analysis of data
    3. Synthesis of data
    4. Testing procedures
C. Conclusion
    1. Summary of findings
    2. Conclusions, development of theory, predictions, and recommendations

The Introduction   The introduction to the thesis informs the reader of its nature and establishes the framework. It describes the area of felt need, defines the problem, and states the possible solutions which will guide the course of study. The justification should modestly present the importance of the study to the discipline of geography. It illuminates the gaps in the knowledge, gaps to be filled by the thesis research.

Because geography is essentially concerned with areal relations, the portion of the earth's surface to be studied should be delimited and the scale of the study given.

In proposing the problem for the study, it is necessary that the writer define any terms which might otherwise be vague to the reader. The selection of the terms to define is left to the judgment of the writer; and certainly, the writer may expect a certain level of competence from the reader of a thesis. However, any word or concept not in normal usage among the anticipated readers, or any word that is used in a context different from the standard, should be carefully defined.

Sometimes it becomes advisable for the writer to clarify limitations of his proposed study. Especially if the title or the problem seems to suggest results more extensive than anticipated. It is not appropriate for the thesis writer to imply, intentionally or not, that he will explain something the final study does not consider. This would be soliciting readers under false pretenses and, perhaps, obtaining tentative committee approval for the research problem as a result of this misunderstanding.

No thesis in geography should be written without knowledge of previous research related to the problem. Investigations have already been conducted are presented both for the protection of the writer and for the edification of the reader. Since a great part of the value of scientific research is in its aid to future scholars, the history and significant trends of prior research, along with the outstanding contributions to the subject, become most important. For the same reason, the major gaps in the literature should be indicated in order to conserve the time of future researchers.

Since the review of the literature is so important, it is not uncommon for the master's thesis to contain a separate chapter on the subject. This represents a thorough review of the geographic works related to the thesis problem and topic, as contrasted to the preliminary survey made for the work plan.

In many theses, a chapter on the organization of the study is included. Such a chapter is considered essential in some disciplines, where method and technique are emphasized more than they are in geography. If this information is not the theme of a separate chapter, it should be given prominent consideration as a chapter subdivision. The writer should explain the methods and techniques employed to organize and test the material. The writer should also outline the process followed in the study in order that the reader may understand exactly why the various steps were taken and in what order they will be encountered.

The Substantive Report   After completing the introductory portion of the thesis, the writer is ready to report the essential findings of the study. In terms of the scientific method, this includes the collection, classification, analysis, synthesis, and testing of the data so that the conclusions of the study may be made evident. This section may be presented in one or several chapters, depending upon the length of the study, the techniques employed, the areas of research, or other factors. If more than one chapter is presented, the relationship of each to the whole must be clear. It is appropriate in this respect to begin an individual chapter with a statement of the purpose of the chapter and to end with a statement of the chapter's contribution to the thesis.

It is apparent that the material embodied in this section of the thesis is the essence of the entire report. In a sense, it *is* the thesis. The other sections of the research first introduce and then conclude the report; but they are peripheral and subordinate to the section of the text referred to here as the substantive report. It is here that the research ability of the potential master scholar is displayed. This is the place to bring into focus the skills of data collecting, the detective instinct, the objectivity, the techniques, and the logical processes of reasoning that have been honed and developed in the preparation of the report.

The Conclusion   The final chapter of the report is a summary of the preceding chapters. It is also appropriate to include suggestions for further studies and to restate, in a more succinct matter, conclusions evident from the previous chapters. However, with the exceptions listed on the next page, no new material should be introduced at this point.

The final chapter should be reserved for the concluding and summarizing remarks which artistically and effectively conclude the study.

Since the summary is a recapitulation of the significant ideas and findings of the thesis, it is an important chapter. The substance recounted should be stated in the words and thoughts of the writer. It is imperative the writer have mastery of the research material if he or she is to be capable of wording the summary well. It is this chapter that a reader might peruse to discover the crux of the study and its significant findings without referring to other sections of the report that present the data and rationale upon which the results were based.

There is one type of "new" material that might be introduced into the final chapter of the thesis. This is a statement of the writer's opinions and conclusions, *based* on the research study. The custom in this regard varies· with geography departments, and the thesis writer should confer with his faculty advisor on this matter. Less questionable is the custom of suggesting future research which might be fruitfully undertaken in related areas of geography.

The Reference Materials   The bibliography is an important part of the thesis or of any research report. It is the format collection of sources related to the research problem, and it should be compiled with care. A "complete" bibliography is one that lists all works the writer has read for the study, regardless of their quality or actual contribution to the research. A "selected" bibliography is usually considered preferable since it is more helpful to future researchers. It lists all works cited or reviewed in the text and all literature that actually contributed to the solution of the problem.

The mechanics of a bibliography for the geographic reports do not differ from bibliographies in other fields. Thus, only a few guidelines are reviewed here. The student is advised to consult a complete manual on thesis writing; his department quite possibly will recommend a specific one. The following are suggestions concerning the bibliography:

1. If more than twenty-five sources are listed, it is advisable to group them under headings such as "books," "periodicals," "newspapers," "government publications," "theses and dissertations," and whatever groupings seem appropriate.
2. The titles in each bibliographic group are alphabetized according to the author's last name or, in the absence of the author's name, the first word of the article.
3. Each entry is single-spaced, with the first line beginning at the left margin, and the following lines indented four spaces.
4. A double space is placed between each entry.
5. The form of all entries must be consistent.

6. Commonly, each bibliographic entry contains the author's name (last name first,) the title of the article and/or book, the place of publication, the publisher, and the date of publication. When scientific notation is used the date follows the author.

An appendix is not required in every thesis. When it is included, it follows the bibliography and is preceded by a single sheet entitled AP-PENDIX. Proper contents of the appendix includes tables, charts, graphs, diagrams, documents, or similar presentations of data relating to the research problem but not conveniently a part of the text. Any pertinent material which is not readily available to the reader may also be included in the appendix. If a large number of entries is included, it is wise to classify and group them as Appendix A, Appendix B, and so forth, with a title for each appendix.

Evaluating the Thesis   The final evaluation is a necessary but, in some ways, an anticlimactic part of the thesis process. If the thesis has been properly conceived, researched, and reported, the writer need have no worry concerning its acceptance. In fact, the oral defense becomes for the master's candidate an opportunity to discuss with others a subject on which he or she has become something of an expert. The thesis is judged on the same points that were itemized in the early stages of planning when the operational plan was designed. There is no general agreement about the emphasis to be placed on the various items of the thesis in the evaluation. There is common agreement, however, on the minimum essentials necessary to meet the quality of the standard thesis. The prerequisites should be checked by the writer carefully before submitting the finished work; for if any of the items are not present, the thesis would probably be unacceptable to the judging and evaluating committee.

The following list of items includes the essential requisites for the thesis. It was compiled by John C. Almack in 1930 and has served six decades as a guide.[2] Although it is called a checking schedule for the thesis, it is not a measuring tool and no values are assigned to the various items.

I. The thesis as a contribution
    A. To knowledge, truth, or
    B. To technique or method or
    C. Knowledge made available not before available
II. The thesis is original
    A. In data or principle or

---

2. John C. Almack, *Research and Thesis Writing* (Boston: Houghton Mifflin Co., 1930), p. 288. Used by permission of the publisher.

B. In technique or method
III. The method is scientific
    A. Normative or
    B. Experimental or
    C. Historical
IV. The results are scientific
    A. A norm or
    B. A law or
    C. History or
    D. New data brought under acceptable principle
V. Requirements of the research process have been met when
    A. There is a problem
    B. There is a hypothesis
    C. The tests of it have been thorough
    D. The source is valid
    E. The data are reliable
VI. The mechanics are correct when
    A. The literature has been reviewed
    B. The introduction is complete
    C. There is a table of contents
    D. There are no typographical or grammatical errors
    E. The charts and tables are in proper form
    F. The conclusion is complete
    G. The bibliography is adequate and in proper form
    H. The form, arrangement, and binding are correct

The Abstract   An abstract of a research report has many uses. It is a capsule account that informs readers about the research, how it was done, and what was accomplished. Abstracts are published in a special compilation for theses, and often they must be submitted before a paper is accepted for presentation at scholarly meetings.

The abstract is composed of three parts which correspond to the three sections of the research text: (1) the statement of problem, (2) the exposition of methods and procedures employed in the study, and (3) a condensed summary of the findings. It is not uncommon to find abstracts composed of three paragraphs that condense the three sections into a few sentences. The length varies, depending upon the research report and its problem, procedures, and findings.

The first part of the abstract describes the problem and the study area. For example, "The objectives of this study are: (1) to describe the pattern of Tennessee voting behavior, (2) to measure the pattern in a manner capable of being duplicated, and (3) to explain the spatial variations of the pattern. The observation units are the ninety-five counties of Tennessee presidential records from 1800 to 1970."

The second section of the abstract defines the manner in which the research has been conducted and the methods, techniques, and tools employed: "Hypotheses are formulated concerning the association between voting patterns (Y) and three independent variables (X) hypothesized as related to the pattern: (1) tradition, (2) race, and (3) income. The test used to determine the distance, degree, and direction of the association among the variables are simple, partial, multiple correlation, and regression analysis."

The final section of the abstract informs the reader of the conclusions of the study. For example, "A highly significant (.01) coefficient of correlation was found to exist between the voting pattern and the independent variables, R = .968. Among the variables, tradition (X_1) explained [r^2] .681 of the "Y" distribution, while race (X_2) and income (X_3) explained [r^2] .421 and .201, respectively. The techniques employed should be applicable to other studies of voting patterns in other states, and results obtained should be directly comparable with those of future studies."

Abstracts may be longer than the above example, and the phraseology will vary in each case.

The Review   The review is a critical appraisal of a book or article and is usually written by a specialist on the topic or area treated in the report. It is written for those who desire to know the content of a work before purchasing or reading it. The student researcher should be familiar with reviews, because *his or her* research reports may be reviewed, or because he or she may be asked to review an article in the future. The reviewer should be objective and should not allow personal preference or bias to influence his judgment. It is the students right, in fact, obligation, to state his or her opinions; but they must be informed opinions and not merely criticism.

Whether the review is long or short, certain essentials should be included. They are: (1) bibliographical information, including the price, (2) the purpose of the work and for whom it is intended, (3) how it is organized, (4) what it says, (5) critical comments, (6) overall value and summary, and (7) the credentials of the author. A hypothetical example utilizing the proper form follows. The numbers are inserted as an aid in identifying the separate parts of the review:

(1) *Population Geography of New England,* Jane Doe. Iowa: Wm. C. Brown Company Publishers, 1971, 200 pp., 49 maps, 20 tables, 40 charts, 21 diagrams, bibliography, $17.50.

(2) This is a short, encyclopedic work published as a part of the School Series in Geography for those teaching population geography and related courses. (3) Written in the traditional style of population geography, the work reflects that the

author has gone to considerable lengths to obtain statistical information about population patterns in the Northeastern United States. (4) Much of the textual material is intended to explain or clarify the many charts and tables presented at a rate of approximately one per page. (5) An objection to the large amount of such material is that much of it is based on 1980 census data which are already out of date. (6) While one might desire a greater emphasis on theoretical framework and more emphasis on analysis, the publication still becomes necessary reading for those seeking a complete understanding of population patterns in New England. (7) Dr. Doe is Professor of Geography at New England University and has previously authored numerous articles on population geography and demography.

Professional and Semi-Professional Articles

Unlike theses and dissertations, professional or semi-professional articles are shorter, usually ranging from two or three pages to ten in length. An article of twenty pages would be unusual. Further, theses and dissertations are written primarily for a faculty committee as a partial requirement for a graduate degree. Normally, the number of readers is small. Journals and magazines, in which articles appear have a much wider distribution, in some cases numbering in the millions. The readers of articles in professional journals have a direct professional interest in the paper. Methodology, analysis, and data acquisition are emphasized, and technical language need not be avoided. Articles in semi-professional journals or magazines are read not only by professional geographers but by non-geographers as well. Descriptions and research results are stressed and details regarding methodology analysis and research design need not be emphasized. Technical jargon should be avoided, and the writing style should be straightforward, and understandable to the non-scientist.

The journals or magazines usually provide the author(s) with explicit instructions regarding style, format, and length. Both the professional and semi-professional articles have several common components. These are: (1) the preliminaries, (2) the text, and (3) the reference materials.

The preliminaries will always include the title of the article, the name of the author or authors, and their affiliations. In addition, some journals will require an abstract and a biographical sketch of the author(s). The text should include: (1) an introduction, (2) substantive report, and (3) a conclusion. The reference materials may be listed at the end of the article or footnoted throughout depending upon the format used by the given journal or magazine. The discussion of these components as they pertain to theses and dissertations are appropriate to professional and semi-professional articles. However, due to length restrictions of articles, and limitations imposed by journals, textual modifications may need to be made to fit the style and format of the specific journal or magazine.[3]

Technical Reports

Normally, technical reports are written for a small, select group of readers who may be supervising or funding a research project and other persons or groups involved in similar research. These types of reports stress the manner in which the research was conducted—personnel involved, funds and time devoted to specific tasks, and types of surveys or experiments conducted. They may be considered as an accounting of how time and funds were expended. Usually, they are written at the completion of the research project and summarize the total research effort. However, in some cases, they may also be prepared at intervals as the research develops to serve as periodic progress reports.

Directions for preparing and writing a technical report range from none to highly detailed, depending upon the group or agency requesting the report. Table 8.1 includes the basic elements that are usually included in a technical report. In lieu of explicit instructions, Table 8.1 could serve as a checklist for the technical report writer. All reports include an introductory section, a substantive section and, if funded, a budgetary section as well. The introductory section typically contains a preface, which provides any necessary background to the project, and an executive summary, which is an overview of one or two pages that highlights the major findings of the project. Most administrators and executives read the summary before anything else because it offers a quick summation needed for those with very busy schedules. Items 1 through 3 in the Substantive Section of Table 8.1 may be a brief statement because the nature and objectives of the research project have previously been approved. Items 4 through 14 of the Substantive Section and all of the Budgetary Section should be as detailed as possible to fully inform the readers about how the research was done, funded, and the results. The Budgetary Section should be explicit about how and when funds were expended, including full justifications.

Textbooks

It is unlikely that a student researcher will undertake the writing of a textbook and, for that reason, little attention will be paid to this type of

---

3. For an excellent review on views for writing for journals, see "Views and Opinions Feature on Writing, Editing and Publishing," *The Professional Geographer*, Vol. 40, No. 1, 1988, pp. 1-18.

TABLE 8.1

Elements of a Technical Report

I. Introduction Section
    1. Preface
    2. Executive Summary
II. Substantive Section
    1. Relevancy of the project
    2. Problem and matrix
    3. Project objectives
    4. Organizational framework
        a) Personnel involved
    5. Tasks completed
    6. Methodologies employed
    7. When tasks were completed
        a) Time schedule of completed research
    8. Results of the research
        a) Conclusion
    9. Impacts of the research
        a) Significance
        b) Who will benefit from the research?
        c) How will they benefit from the research?
    10. Evaluation of the project
        a) Methodology employed
        b) Time span of the evaluation
        c) Personnel who conducted the evaluation
        d) Results and conclusions
    11. Innovations
        a) New methodologies, techniques, or surveys developed and employed?
    12. Dissemination of results
        a) Methodology employed
        b) Personnel who conducted dissemination
        c) Time span of dissemination
        d) Feedback
    13. Proposed future activities
        a) New avenues of research
    14. Appendices
        a) Time Charts
        b) Instruments utilized
        c) Description of surveys/experiments
III. Budgetary Section
    1. Personnel
        a) Principal investigator
        b) Research associates/assistants
        c) Secretaries/clerical help
        d) Consultants
        e) Fringe benefits
    2. Travel
        a) Travel stipends—domestic and foreign

b) Subsistence/per diem
3. Other direct costs
   a) Permanent equipment
   b) Computer time—rental
   c) Materials and supplies—non-expendable
   d) Publication costs
4. Indirect costs
   a) Phone
   b) Rent
   c) Utilities
   d) Materials and supplies—expendable
   e) Other—specify

---

geographic writing here. Texts are normally written by geographers with many years of experience as researchers, authors of professional papers, and teachers. Textbooks vary greatly in subject matter content and the manner in which the research work is conducted. For example, a text in World Regional Geography or Introduction to Physical Geography is researched using secondary and tertiary source materials for the most part. On the other hand, a text or monography entitled Advanced Climatology or Cultural Diffusion in the Balkans probably would deal with primary source materials to a greater extent.

The author(s) of textbooks are normally responsible for the subject matter content and the reference materials. The editors of the publishing company organize the preliminary materials, and control the style, format, and design of the book. Typographical errors and grammatical mistakes may be corrected prior to the publishing of the second and later printings. However, no major changes can be made at this time. Major revisions, additions, and deletions can be made when a second or later editions are being prepared and written.

Term and Seminar Papers

It is common practice in American colleges and universities to require term papers in upper level undergraduate courses, and most senior or graduate seminars require a written research report of some sort. It is the rare geography student who has not written one or more term papers at some point in his or her undergraduate or graduate career. The term paper is similar to a master's thesis in that it is an individual effort. The thesis is normally longer and represents a more intensive research effort extended over a longer period of time. Normally a term paper involves library research only and does not have a problem to be addressed or an hypothesis to test. It has a general topic or theme related to the subject matter content of the course or class for which it is being prepared.

In most cases, guidelines are provided by the instructor regarding length limitations, organization, and format. Most term or seminar papers include:

1. Title page
2. Table of contents
3. List of tables (if appropriate)
4. List of illustrations (if appropriate)
5. Text
6. Bibliography
7. Appendixes (if appropriate)

The title page should include the title of the paper, the student's full name, and the date (month and year) that the paper is submitted. The title of the course for which the paper has been written may also be included on the title page. The paper may not have chapters as such, but there will be major headings, separating the textual material into orderly divisions. These major divisions of the paper should be listed in the table of contents. If there is more than one table in the paper, a list of tables would be appropriate. The same holds true for a list of illustrations. Photographs, maps, diagrams, charts, and sketches may all be considered illustrations and numbered in order as they appear in the paper. Term papers normally do not have an index. Therefore, the readers must rely heavily on the table of contents, list of tables, and list of illustrations to locate specific sections of the paper.

The text of the term paper is organized in the same manner as the master's thesis. There is an introduction, substantive section, and a conclusion. The introduction should state the justification, need, and relevancy of the study, references to pertinent literature, and the organization of the paper. The substantive section is the heart of the paper and all salient information and major points should be clearly and forcibly discussed at this point. The conclusion may be a summary of the paper's high points. No new material should be introduced at this time.

The bibliography is an important part of the term paper as it is a list of published sources that directly relate to the subject matter of the paper. A selected bibliography is preferable for a term paper, listing only sources cited or actually reviewed. Normally, the magnitude and scope of term or seminar papers are such that appendixes are not part of the report. Term papers are concerned with a general theme and usually are relatively short in length. In some unusual cases, however, an appendix may be appropriate. For example, a table illustrating the demographic and economic statistics of a given state or country, or a survey instrument such as a questionnaire, or an elaborate chart may be included in the paper as an appendix providing that it bears directly on the subject matter of the paper but is too cumbersome to be included in the textual portion of the report. If an appendix is included, it must be referred to in the textual portion of the paper.

Proposals

Proposals are prepared and written for one of two reasons: (1) to seek approval or permission from a person or group to undertake a proposed research project or begin a specific type of investigation, and (2) to obtain funds to support a research project. In both situations, the overriding object is to present the strongest possible case to justify the proposed research endeavor. An example of the first case might be a student researcher requesting approval from a faculty member or committee regarding a master's thesis or doctoral dissertation. In such cases, the academic department or college usually provides a standardized form that the student fills out and submits to the proper authority. The complete form contains all the necessary information, and essentially serves as the proposal. However, in many cases, there may be no guidelines to follow in preparing proposals for financial support. Directives range from none to detailed, depending upon the policies of a given foundation. Most researchers, at one time or another, will feel the need to obtain additional funds for a part or all of a proposed research project. A student researcher may wish to obtain funds to offset the costs of researching a thesis or dissertation. Writing proposals is time-consuming but often necessary for many types of geographic research carried on today. The competition for research funding is keen, because normally there are requests for more money than existing available funds.

In addition to obtaining funds, writing proposals has another important aspect. The writing of the proposal forces the researchers to think through the entire research project, determine clearly what is to be done, organize their resources, justify their methods, and anticipate the impact of the project when completed. This, in itself, is a major part of the overall research process. In fact, some researchers write proposals primarily with this goal in mind and with the acquisition of funds being of secondary concern.

Although the required or suggested format and content of proposals will vary somewhat from one funding group to another, there are several common steps and major components in proposal construction. The emphasis of each component will vary, and two or more may be combined, depending upon the nature of a specific proposal and the directives of the foundation or funding agency. These steps and components are:

1. Initial investigation

2. Cover letter
3. Summary of the proposal
4. Proposal introduction
5. Problem statement and/or needs assessment
6. Problem objectives
7. Project design and methodologies to be employed
8. Commitment of personal and institution
9. Personnel qualifications
10. Evaluation plan
11. Potential impact
12. Budget

Initial Investigation   The obvious first step in proposal construction is to determine whether or not a proposed research project is compatible with the objectives of a given funding agency. A careful study of all brochures, announcements, and correspondence must be undertaken to determine if the researchers, the institutions, and the research discipline are eligible for consideration by this particular funding group or program. Further, the proposed project must fit the fiscal and/or time limitations set by the funding source. If this investigation indicates that the proposed project is indeed eligible and compatible with the funding group's stated purpose and scope, the actual writing of the proposal can then be undertaken.

Cover Letter   Proposals are usually accompanied by a cover letter. The purpose of this letter is to assure the funding source that the authorities or administrative officers of the researcher's institution endorse and support the proposed project. It should *briefly describe* the proposal, include the names of the proposed investigators, amount of money requested, desired starting date and duration of the proposed project, and addresses and phone numbers of principal directors or investigators.[4]

Summary   The summary of the proposal should be clear, concise, and written so that a layperson can understand the problem, the need, and the use of funds to support the project. The kinds of activities to be carried on and the goals to be achieved should be stressed. The summary should be written with great care because it is the first part of the proposal to be read and often serves as a statement for public distribution at a later date. Although the summary normally appears in the beginning of the proposal, it can be best written after the entire proposal has been prepared.

---

4. An excellent paper describing proposal writing is Kiritz, N. J.: "Program Planning and Proposal Writing," *The Grantmanship Center News*, Los Angeles, CA, May/June 1979. Also see special section "Directions In Geography: Feature on Funded Research," *The Professional Geographer*, Vol. 41, No. 1, 1989, pp. 2-10.

Introduction   The proposal introduction should include a brief statement describing the purpose, goals, and philosophy of the researcher's institution; how and when it became established; its qualifications and facilities for conducting the proposed research; and other related research projects or programs currently carried or completed recently. All credibility-related data concerning the institution or department that is directly appropriate to the proposed research should be included in this part of the proposal.

Problem Statement   The problem statement is the core of the proposal and the *raison d'être*. It should clearly state what the problem is and why it needs to be studied. A documented review of current and past activities in this area is highly desirable, supported by statistical data and authoritative statements. The context of the proposed project as it relates to the current and past activities of the institution should be explained as well as its relationships to the local and national situation, if appropriate. The other components of the proposal are essentially supplementary and provide supportive details to this section. The problem statement, including the urgency of solving the problem and meeting what needs currently exist, is the major overriding part of the proposal.

Project Objectives   The section concerned with the project objectives should illustrate how the project will contribute to currently important scientific developments or concerns. The anticipated outcomes should be described in detail and include what population will derive what kinds of benefits and when. Statistical documentation is desirable, if at all possible.

Project Design   The project design should describe in detail how the research is to be carried out. The methodologies and techniques to be employed should be discussed, and work plans, timetables, and milestones for project activities indicated. The activities of the project personnel, and the organization and management framework should be described. Tables and charts, to provide a clear view of the project activities, often are most helpful. Program or project planning that describes the sequence of tasks and how each major task is to be accomplished is also desirable.

Personnel Commitment   A detailed statement that elaborates the points that were briefly discussed in the cover letter about the commitment to the project by the researcher's institution and project personnel is often a necessary part of the proposal. It is important to assure the funding group that the institution and project personnel are enthusiastic and, if funded, will devote their efforts and facilities to the completion of the project.

Personnel Qualifications   The qualifications of each professional person with responsibilities in the proposed project should be described. This would include their experience and education directly pertinent to the proposed project activities. Normally, a curriculum vitae or resume for each of the project's research persons would be included in an appendix.

Evaluation Plan   A plan to assess or evaluate the success of the project is necessary, and this plan may also serve as a mechanism to obtain feedback for the improvement and further development of the project. The plan should clearly state who will conduct the evaluation and how the evaluation process will be conducted. The manner in which evaluation criteria are defined, and the way in which the evaluation data are obtained and processed are the critical components of the overall evaluation plan. Each research project is unique to some degree and the evaluation procedure often must be tailored to fit the specific project.

Potential Impact   The project, when completed, will have some kind of an impact on a given population either locally or regionally or perhaps nationally. This potential impact should be discussed as specifically as possible. Further, the results of the project might lead to other research tasks or continuation of the project. The potential impacts and plans for future projects should be stated as specifically as possible to provide the funding source with a long range plan if the project proves to be highly successful.

Budget   Each funding source normally has detailed instructions concerning budgeting matters. The standard format usually includes personnel costs, non-personnel costs, and indirect costs. Often a budget summary is required and is included as the first page or pages of the budget section. Preparing the budget must be done with great care, closely adhering to the funding source's instructions and limitations. In all cases, the proposal writer must be specific and justify all expenditures compatible with the guidelines of the funding agency and the researcher's institution.

In the last analysis, a good proposal clearly states what it is you intend to do, why it is important, and appropriateness of the methodologies to be employed, the qualifications of the personnel, the facilities to be used, the potential impact, and the precise cost. In the great majority of cases, such a proposal that includes these components will provide the answers to the questions of the proposal reviewers.

Summary

Writing research reports is an integral and important part of the research process. The need to record research is not only for the benefit of the researcher, but is of value to others as well. Therefore, it is important to communicate with the reader in the most effective manner. It has been said that written communication is an art as well as a science, and the first step to success is to identify the main body of potential readers. These readers may be authorities on the subject in which the research was conducted, such as a faculty committee supervising an academic thesis or dissertation. On the other hand, the readers may have only a cursory or passing interest in an article published in a semi-professional magazine. In writing about the results of a research project, the researcher must always bear in mind the background and interest level of the main body of prospective readers. The style of writing, type of illustrations, and the organization of major points should be tailored, to the greatest degree possible, for this particular audience.

Selected References

·

The following list of references and source materials is not intended to be complete. A comprehensive listing would be so large as to be impractical for a book of this nature. The listing does, however, include many of the more widely used references in geographic research.

I. GENERAL REFERENCES—THE NATURE AND SCOPE OF MODERN GEOGRAPHY

ABLER, R.; ADAMS, J. S.; and GOULD, P. *Spatial Organization: The Geographer's View of the World.* Englewood Cliffs, N.J.: Prentice-Hall, 1971.

ACKERMAN, EDWARD A. *Geography as a Fundamental Research Discipline.* Department of Geography Research Paper No. 53. Chicago: University of Chicago Press, 1958.

———., ed. *The Science of Geography.* Washington, D.C.: National Academy of Science-Natural Research Council, 1965.

AMEDEO, D. and GOLLEGE, R. G., *An Introduction to Scientific Reasoning in Geography.* New York: John Wiley & Sons, 1975.

BREWER, J. B. *The Literature of Geography.* Hamden, Conn.: Linnet Books, 1973.

BROEK, JAN O. M. *Compass of Geography*. Columbus, Ohio: Charles E. Merrill Publishing Co., 1966.

BUTTIMER, A. *Values in Geography*. Commission on College Geography, Resource Paper No. 24. Washington, D.C.: Association of American Geographers.

BUNGE, W., *Theoretical Geography,* Lund Studies in Geography. Lund, Sweden: The Royal University of Lund, C. W. K. Gleerup, 1966.

CHORLEY, R. J. and HAGGETT, P., eds., *Models in Geography*. Toronto, Ontario: Methuen Publications, 1967.

COHEN, SAUL B., ed., *Problems and Trends in American Geography*. New York: Basic Books, 1967.

COOKE, R. U. and JOHNSON, J. H. *Trends in Geography: An Introductory Survey*. London: Pergamon Press, 1969.

DAUGHERTY, RICHARD, *Science in Geography: Data Collection*. London: Oxford University Press, 1974.

FIELDING, G. J. *Geography as Social Science*. New York: Harper & Row, Publishers, 1974.

FITZGERARD, B P., *Science in Geography: Development in Geographical Method*. London: Oxford Press, 1974.

FRAZIER, J. W., ed. *Applied Geography: Selected Perspectives*. Englewood Cliffs: Prentice-Hall, 1982.

FREEMAN, T. W. *A Hundred Years of Geography*. Chicago: Aldine Publishing Co., 1961.

FUSON, ROBERT H. *A Geography of Geography*. Dubuque, Iowa: Wm. C. Brown Company Publishers, 1969.

Geography and Geography. Anglo-American Human Geography since 1945. London: Edward Arnold, Ltd., 1983.

HAGGETT, PETER. *Geography: A Modern Synthesis*. New York: Harper & Row, Publishers, 1972.

HAGGETT, PETER. *Locational Analysis in Geography*. New York: St. Martin's Press, 1965.

HARTSHORNE, RICHARD. *Perspective on the Nature of Geography*. Chicago: Rand McNally & Co. (for the Association of American Geographers), 1959.

HARVEY, D. W, *Explanation in Geography*. New York: The Natural History Press, 1970.

INHABER, H. *Environmental Indices*. New York: John Wiley & Sons, 1976.

JAMES, P. E. *All Possible Worlds: A History of Geographical Ideas*. New York: Bobbs-Merrill Co., 1972.

JAMES, P. E. and JONES, C. F., eds. *American Geography: Inventory and Prospect*. Syracuse, N.Y.: Syracuse University Press, 1954.

JOHNSTON, R. J. *Philosophy and Human Geography. An Introduction to Contemporary Approaches.* London: Edward Arnold, Ltd., 1983.

LOUNSBURY, J. F., ed. *Land Use: A Spatial Approach.* Dubuque: Kendall/Hunt Publishing Co., 1981.

McCULLAGH, PATRICK. *Science in Geography: Data Use and Interpretation.* London: Oxford University Press, 1974.

SCHMEIDER, A. A.; GRIFFIN, P. F.; CHATHAM, R. L.; and NATOLI, S. J. *A Directory of Basic Geography.* Boston: Allyn & Bacon, 1970.

TAAFFE, EDWARD J., ed. *Geography.* Englewood Cliffs, N.J.: Prentice-Hall, 1970.

TAYLOR, GRIFFITH, ed. *Geography in the Twentieth-Century.* New York: Philosophical Library, 1957.

WILMOTT, C. J. and GAILE, G. L. *Geography in America.* Columbus, Ohio: Merrill Publishing Co., 1989.

WOOLDRIDGE, S. W. and EAST, W. G. *The Spirit and Purpose of Geography.* New York: G. P. Putman's Sons, 1967.

WRIGHT, JOHN K. *Human Nature in Geography.* Cambridge, Mass.: Harvard University Press, 1966.

II. SPECIFIC REFERENCES—RESEARCH METHODS
  A. Field and Cartographic Methods, Remote Sensing

AVERY, T. E. and GRAYDON, L. B. *Interpretation of Aerial Photographs.* Minneapolis: Burgess Publishing Company, 1985.

BIRCH, T. W *Maps: Topographical and Statistical.* London: Oxford University Press, 1964.

CAMPBELL, J. *Map Use and Analysis.* Dubuque, IA: Wm. C. Brown Publishers, 1989.

DICKINSON, G. C., *Maps and Air Photographs,* London: Edward Arnold, 1969.

DOWNS, R. M. and STEA, D. *Maps in Mind: Reflections on Cognitive Mapping.* New York: Harper & Row, Publishers, 1977.

ESTES, J. E. and SENGER, L. W., eds. *Remote Sensing: Techniques for Environmental Analysis.* Santa Barbara, California: Hamilton Publishing Co., 1974.

FRIBERG, J. C. *Field Techniques and the Training of the American Geographer,* Discussion Paper No. 5, Department of Geography, Syracuse University, 1975.

GREENHOOD, DAVID. *Mapping.* Phoenix Science Series. Chicago: University of Chicago Press, 1964.

GUNN, ANGUS M. *Techniques in Field Geography.* Toronto, Ont.: Copp Clark Pub. Co., 1962.

HART, JOHN FRASER, ed. *Field Training in Geography*. Commission on College Geography, Technical Paper No. 1, Washington, D.C.: Association of American Geographers, 1968.

HOLZ, R. K., ed. *The Surveillant Science: Remote Sensing of the Environment*. Boston: Houghton Mifflin Company, 1973.

JONES, C. F., and PICO, R., eds. *Symposium on the Geography of Puerto Rico*: University of Puerto Rico Press, 1955.

LILLESAND, T. M. and KIEFER, R. W. *Remote Sensing and Image Interpretation*. New York: John Wiley and Sons, 1987.

LOW, JULIAN W. *Plane Table Mapping*. New York: Harper & Row, Publishers, 1952.

LOUNSBURY, J. F. and ALDRICH, F. T., *Introduction to Geographic Field Methods and Techniques*. Columbus: Charles E. Merrill Publishing Co., 1986.

Manual of Photographic Interpretation. Washington, D.C.: American Society of Photogrammetry, 1960.

Manual of Photographic Interpretation. Washington, D.C.: American Society of Photogrammetry, 1961.

Map Reading. United States Department of the Army, Technical Manual FM 21-26, 1969.

MILLER, V. C. and MILLER, C. F. *Photography*. New York: McGraw-Hill Book Co., 1961.

MONKHOUSE, F. J. and WILKINSON, H. R. *Maps and Diagrams*. New York: E. P. Dutton & Co., 1963.

MONMONIER, M. S. *Computed Assisted Cartography: Principles and Prospects*. Englewood Cliffs: Prentice-Hall, 1982.

MUEHRCKE, PHILLIP. *Thematic Cartography*. Commission on College Geography, Resource Paper No. 19, Washington, D.C.: Association of American Geographers, 1972.

PEUCKER, T. K. *Computer Cartography*. Commission on College Geography, Resource Paper No. 17, Washington, D.C.: Association of American Geographers, 1972.

PLATT, ROBERT S. *Field Study in American Geography*. Department of Geography Research Paper No. 61. Chicago: University of Chicago Press, 1959.

PUGH, J. C., *Surveying for Field Scientists*. Pittsburgh, Pennsylvania: University of Pittsburgh Press, 1975.

RICHASON, B. F. Jr., Ed. *Introduction to Remote Sensing of the Environment*. Dubuque: Kendall/Hunt Publishing Co., 1978.

ROBINSON, ARTHUR H. *The Look of Maps: An Examination of Cartographic Design*. Madison, Wis.: University of Wisconsin Press, 1952.

Rural Land Classification Program of Puerto Rico, The. Northwestern University Studies in Geography No. 1. Evanston, Ill.: Northwestern University Press, 1952.

SPEAK, P. and CARTER, A. H. C. *Map Reading and Interpretation*. London: Longmans Group Ltd., 1977.

STODDARD, R. H. *Field Techniques and Research Methods in Geography*. Dubuque: Kendall/Hunt Publishing Co., 1982.

WHEELER, K. S. and HARDING, M. *Geographical Field Work*. London: Anthony Blond, 1965.

B. Statistical Methods, Sampling

BERRY, B. J. L. *Land Use, Urban Form and Environmental Quality*. Department of Geography Research Paper No. 155. Chicago: University of Chicago Press, 1974.

BERRY, B. J. L., and MARBLE, D. F., eds. *Spatial Analysis: A Reader in Statistical Geography*. Englewood Cliffs, N.J.: Prentice-Hall, 1968.

CHORLEY, R. J., and HAGGETT, P., eds. *Models in Geography*. Toronto, Ont.: Methuen Publications, 1967.

COCHRAN, W. G. *Sampling Techniques*. New York: John Wiley & Sons, 1963.

COLE, J. P., and KING, C. A. M. *Quantitative Geography*. New York: John Wiley & Sons, 1968.

GREER-WOOTTEN, B. *A Bibliography of Statistical Applications in Geography*, Commission on College Geography, Technical Paper No. 9. Washington, D.C.: Association of American Geographers, 1972.

GREGORY, S. *Statistical Methods and the Geographer*. London: Longmans Group, Ltd., 1968.

GRIFFITH, D., and AMRHEIN, C. G. *Statistical Analysis for Geographers*. Englewood Cliffs, N.J.: Prentice-Hall, Inc., 1991.

HAMMOND, R., and McCULLACH, P. *Quantitative Techniques in Geography: An Introduction*. London: Oxford University Press, 1974.

HARVEY, D. W. "Data Collection and Representation in Geography." *Explanation in Geography*, 266–81. New York: The Natural History Press, 1970.

KING, L. J. *Statistical Analysis in Geography*. Englewood Cliffs, N.J.: Prentice-Hall, 1969.

KRUMBEIN, W. C. and GRAYBILL, F. A. *An Introduction to Statistical Models in Geology*. New York: McGraw-Hill Book Co., 1965.

KUCHLER, A. W. *Vegetation Mapping*. New York: Ronald Press Co., 1967.

MATALAS, N. C. "Geographic Sampling." *The Geographical Review*, Vol. 4, 606–608. New York: The American Geographical Society of New York, 1963.

SMITH, R. H. T.; TAAFFE, E. J., and KING, L. J. *Readings in Economic Geography: The Location of Economic Activity*. Chicago: Rand McNally & Co., 1968.

STUART, A. *Basic Ideas of Scientific Sampling*. New York: Hafner Press, 1962.

THEAKSTONE, W. H. and HARRISON, C. *The Analysis of Geographical Data*. London: Heinemann Educational Books, 1970.

WOOD, W. F. "Use of Stratified Random Sample in a Land Use Study," *Annals of the Association of American Geographers*, Vol. 45, 350–367. Washington, D.C.: The Association of American Geographers, 1955.

YEATES, MAURICE H. *An Introduction to Quantitative Analysis in Economic Geography*. New York: McGraw-Hill Book Co., 1968.

C. Interviewing and Questionnaires

BABBIE, E. R. *The Practice of Social Research*. Belmont, Calif.: Wadsworth Publishing Co., 1975.

DAVIS, J. A. *Elementary Survey Analysis*. Englewood Cliffs, N.J.: Prentice-Hall, 1971.

GORDEN, R. L. *Interviewing: Strategy, Techniques, and Tactics*. Homewood, Ill.: Dorsey Press, 1969.

HIGHSMITH, R. M. JR. "Suggestions for Improving Geographical Interview Techniques," *Professional Geographer*, January, 53–55. Washington, D.C.: Association of American Geographers, 1962.

KAHN, R. L. and CANNELL, C. F. *The Dynamics of Interviewing*. New York: John Wiley & Sons, 1967.

KNIFFEN, F. "The Tape Recorder in Field Research." *Professional Geographer*, January, 83. Washington, D.C.: Association of American Geographers, 1962.

MILLER, D. C. *Handbook of Research Design and Social Measurement*. New York: Longman, 1983.

OPPENHEIM, A. W. *Questionnaire Design and Attitude Measurement*. New York: Basic Books, 1966.

PAYNE, S. L. *The Art of Asking Questions*. Princeton, N.J.: Princeton University Press. 1951.

RICHARDSON, S. A.; DOHRENWEND, B. S.; and KLEIN, D. *Interviewing: Its Form and Functions.* New York: Basic Books, 1965.

SHESKIN, I. M. *Survey Research for Geographers.* Washington, D.C.: Association of American Geographers, 1985.

WARWICK, D. P. and LININGER, C. A. *The Sample Survey: Theory and Practice.* New York: McGraw-Hill, 1975.

WARWICK, D. P. and OSHERSON S. *Comparative Research Methods.* Englewood Cliffs, N.J.: Prentice-Hall, 1973.

III. SELECTED GEOGRAPHICAL PERIODICALS AND SERIALS
    A. United States and Canada

Annals of the Association of American Geographers. Association of American Geographers, 1710 Sixteenth Street, N.W., Washington, D.C. 20009.

California Geographer, The. Journal of the California Council for Geography Teachers, Department of Geography, California State College, Long Beach, Calif. 90801.

Canadian Geographer, The. The University of Toronto, Toronto, Ontario, Canada.

Cartography and Geography Information Systems. American Congress of Surveying and Mapping, Bethesda, Maryland.

Commission on College Geography. (Regular Series; Resource Papers; Technical Papers). Association of American Geographers, 1710 Sixteenth Street N.W., Washington, D.C. 20009.

Department of Geography Research Papers. Department of Geography, University of Chicago, 1101 East 58th Street, Chicago, Ill. 60637.

Discussion Paper Series. Department of Geography, University of Iowa, Iowa City, Iowa 52240.

Discussion Papers. Department of Geography, The Ohio State University, 1775 South College Road, Columbus, Ohio 43210.

East Lakes Geographer, The. Journal of the East Lakes Division, Association of American Geographers, Department of Geography, 1775 South College Road, Ohio State University, Columbus, Ohio 43210.

Economic Geography. Department of Geography, Clark University, Worcester, Mass. 01610.

Focus. The American Geographical Society of New York, Broadway at 156th Street, New York, N.Y. 10032.

Geographical Analysis. Columbus, Ohio: The Ohio State University Press.

Geographical Bulletin, The. Geographical Branch, Department of Mines and Technical Surveys, Ottawa, Canada.

Geographe Canadien (Canadian Geographer). Canadian Association of Geographers, Morrice Hall, McGill University, Montreal 2, P. Q., Canada.

Geographical Review, The. The American Geographical Society of New York, Broadway at 156th Street, New York, N.Y. 10032.

Journal of Geography, The. National Council for Geographic Education. Western Illinois University, Macomb, Illinois 61455.

Journal of Historical Geography. Academic Press, New York.

Monograph Series of the Association of American Geographers. The Association of American Geographers, 1710 Sixteenth Street, N.W., Washington, D.C. 20009.

National Geographic Magazine. The National Geographic Society, Seventeenth and M. Streets, N.W., Washington, D.C. 20036.

Papers and Proceedings of Applied Geography Conferences. Department of Geography, SUNY-Binghamton, Binghamton, N.Y. 13901

Physical Geography. V. H. Winston & Son, Inc., Silver Spring, Maryland.

Professional Geographer, The. Forum and Journal of the Association of American Geographers, 1710 Sixteenth Street, N.W., Washington, D.C. 20036.

Professional Papers. Department of Geography and Geology. Terre Haute, Ind.: Indiana State University.

Southeastern Geographer, The. Journal of the Southeastern Division, Association of American Geographers, Department of Geography, University of North Carolina, Chapel Hill, N.C. 27514.

Soviet Geography: Review and Translation. The American Geographical Society of New York, Broadway at 156th Street, New York, N.Y. 10032.

Studies in Geography. Department of Geography, Northwestern University, Evanston, Ill. 60201.

Urban Geography. V. H. Winston & Son, Inc., Silver Spring, Maryland.

B. Others

Antipode: A Radical Journal of Geography. Basil Blackwell, Oxford, Cambridge, Mass.

Applied Geography. Butterworth-Heinemann, Ltd., Oxford, U.K.

Australian Geographer, The. The Geographical Society of New South Wales, Department of Geography, University College, Newcastle, N. S. W., Australia.

Australian Geographical Studies. Journal of the Institute of Australian Geographers, Department of Geography, University of Melbourne, Parkville, N. 2, Victoria, Australia.

Geographical Journal, The. The Royal Geographical Society, London, S. W. 7, England.

Geographical Magazine, The. Geographical Magazine, Ltd., Friars Bridge House, Queen Victoria Street, London, E. C. 4, England.

Geography. George Philip & Son, 32 Fleet Street, London E. C. 4, England.

Institute of British Geographers. Transaction New Series. Basil Blackwell, London, U.K.

International Journal of Geographical Information Systems. Taylor and Francis, London, U.K. and Washington, D.C., USA.

New Zealand Geographer. The New Zealand Geographical Society, Department of Geography, University of Canterbury. Christchurch, New Zealand.

Political Geography Quarterly. Butterworth-Heinemann, Ltd., Oxford, U.K.

Publications of the Institute of British Geographers. The Institute of British Geographers, Department of Geography, Cambridge University, Cambridge, England.

Scottish Geographical Magazine, The. The Royal Scottish Geographical Society, Synod Hall, Castle Terrace, Edinburgh, Scotland.

NOTE: The previous list of periodicals and serials is selective and includes only American, Canadian, Australian, United Kingdom and New Zealand publications. For a comprehensive listing of geographical periodicals in English, including those of other countries, the research student is advised to refer to:

HARRIS, C. D. and FELLMANN, J. D. *International List of Geographical Series,* 3rd ed., Department of Geography Research Paper No. 193. Chicago: University of Chicago Press, 1980.

HARRIS, C. D. *Guide to Geographical Bibliographies and Reference Works in Russian or on the Soviet Union.* Department of Geography Research Paper No. 164. Chicago: University of Chicago Press, 1975.

IV. BIBLIOGRAPHIC REFERENCES

BROWN, C. L. and J. O. WHEELER. *A Bibliography of Geographic Thought.* New York: Greenwood Press, 1989.

Current Geographical Publications. The American Geographical Society of New York, Broadway at 156th Street, New York, N.Y. Monthly, except July and August.

DURRENBERGER, ROBERT W. *Environment of Man: A Bibliography.* National Press Books, Palo Alto, Calif., 1970.

Geographical Bibliography for American Colleges, A. Commission on College Geography, Publication No. 9, Association of American Geographers, 1710 Sixteenth Street, N.W., Washington, D.C. 20009, 1970.

HARRIS, C. D. (editor). *A Geographical Bibliography for American Libraries.* Washington, D.C.: Association of American Geographers, 1985.

HARRIS, CHAUNCY, D. *Annotated World List of Selected Current Geographical Serials.* 4th ed., Department of Geography Research Paper No. 194, University of Chicago Press, 1980.

HORNSTEIN, HUGH, A. *A Bibliography of Paperback Books Relating to Geography.* NCGE Bibliography Series, National Council for Geographic Education, Chicago, Ill., 1970.

Inexpensive Books in Physical Geography. Mimeographed. Compiled by Harold A. Winters, Panel on Physical Geography, Commission on College Geography, 1710 Sixteenth Street, N.W., Washington, D.C. 20009, 1970.

VINGE, C. L. and VINGE, A. G. *U.S. Government Publications for Research and Teaching in Geography and Related Social and Natural Sciences.* Totowa, N.J.: Littlefield, Adams & Co., 1967.

V. STYLE MANUALS AND WRITING GUIDES

CAMPBELL, WILLIAM G. *Form and Style in Thesis Writing.* 3rd. ed. Boston: Houghton Mifflin Co., 1969.

DEIGHTON, L. C. *Handbook of American English Spelling.* New York: Harcourt, Brace, Jovanovich, 1973.

DURRENBERGER, ROBERT W. *Geographical Research and Writing.* New York: Thomas Y. Crowell Co., 1971.

"Editorial Policy Statement." *Annals of the Association of American Geographers,* vol. 60, March, 1970, pp. 194–207. Washington, D.C.: Association of American Geographers, 1970.

MLA Style Sheet, The. Rev. ed. Compiled by William Riley Parker. New York: The Modern Language Association of America, 1968.

Publication Manual. American Psychological Association, Washington, D.C.: American Psychological Association, 1974.

TURABIAN, KATE L. *A Manual for Writers of Term Papers, Theses, and Dissertations.* 4th ed. Chicago: University of Chicago Press, 1973.

Webster's Instant Word Guide. Springfield, Mass.: A and C. Merriam Co., 1972.

Generalized Classification of the
Physical Characteristics of the Land

.

I n the event that detailed data concerning the physical nature of the land is desired, the following classification system may be used for multifeatured or fractional code mapping on aerial photographs or other types of base maps. The student researcher may wish to modify this classification to "fit" his research area better.

    I. *First Symbol*—Slope
        1. 0° to 3° (level to gently sloping).
        2. 4° to 8° (undulating to rolling).

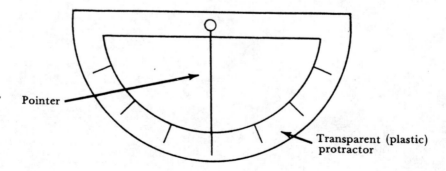

Pointer

Transparent (plastic)
protractor

3. 9° to 14° (rolling to hilly).
4. 15° to 20° (hilly to steep slopes).
5. 21° to 30° (steep slopes).
6. 31° to more (very steep slopes).

> Note: Slopes may be determined and mapped by using a large protractor with a free swinging metal pointer (straighten paper clip, hat pin, etc.).

II. *Second Symbol*—Drainage
   1. *Excessively drained land.* Land which, in terms of drainage, does not retain sufficient moisture for agricultural purposes.
   2. *Adequately drained land.* Land which, in terms of drainage, does not normally present problems of cultivation or agricultural production, but which under heavy rainfall conditions, may necessitate delays in farm operations and may result in reduced crop yields.
   3. *Poorly drained land.* Land which, in terms of drainage, presents problems of cultivation and agriculture production, even during conditions of normal rainfall.
   4. *Very poorly drained land.* Land which, in terms of drainage, cannot be used for agricultural purposes. Virtually swampland.

III. *Third Symbol*—Erosion
   1. *Little or no observable sheet erosion.* Average removal less than 25 percent of the topsoil; if the original topsoil was about 16 inches, the present topsoil should be 12 or more inches.
   2. *Moderate sheet erosion.* Average removal of from 25 to 75 percent of the topsoil; that is, if the original topsoil was about 16 inches, the present topsoil is between 4 and 12 inches.
   3. *Incipient gullying.* Occasional gullies; in general, more than 100 feet apart; shallow, so that they can be crossed by tillage implements but gullies would not be obliterated by normal tillage.
   4. *Damaging gully erosion.* Gullies occurring less than 100 feet apart, but not including more than 50 percent of the total mapping unit; at least one major gully not crossable with tillage implements.
   5. *Excessive gully erosion.* Conditions in excess of those in #4 above. Virtually precludes cultivation.

VI. *Fourth Symbol*—Surface Stoniness and Rock Exposure

1. *Free from stones.* Surface entirely free from stones, or sufficiently free from stones so that there are no difficulties in cultivating the land.

2. *Moderately stony.* Small-sized or large-sized stones at the surface in sufficient quantity to impede cultivation or render it difficult.

3. *Very stony.* Sufficient quantity or small- or large-sized stones to virtually preclude cultivation.

4. *Bedrock exposure.* Bedrock exposed, or is within plow depth of the surface and covers sufficient space to hinder severely or preclude cultivation.

V. *Fifth Symbol*—Soil Type

1. *Bottomland soils.* Clay to silt topsoil, subject to flooding.

2. *Bottomland soils.* Sand to gravel topsoil, subject to flooding.

3. *Terrace soils.* Clay to silt topsoil, normally not subject to flooding.

4. *Terrace soils.* Sand to gravel topsoil, normally not subject to flooding.

5. *Upland soils.* Clay to silt topsoil.

6. *Upland soils.* Sand to gravel topsoil.

Note: The classification of soils above is highly generalized. The researcher is strongly advised to obtain pertinent county soil surveys developed by the Soil Conservation Service, if they are available, and modify his classification accordingly.

Classification of Land Use

This classification, presented here in abbreviated form, is designed to enable the researcher to collect and compile land-use data by using either field mapping or aerial photo interpretation techniques. It is a stratified classification, and data may be collected in varying degrees of detail compatible to the scale of mapping. The researcher may add, or otherwise modify within the existing format, to suit his particular needs without difficulty.

First Symbol—Major Land Use

R. Rural Land   U. Urban Land

Second Symbol—General Land Use

R. Rural Land Use

1. Cropped Land	2. Pasture Land	3. Forest, Grassland, Shrub	4. Idle Land	5. Miscellaneous

Third Symbol—Specific Land Use[1]

1. Cropped Land	2. Pasture Land[2]	3. Forest, Grassland, Shrub[3]	4. Idle Land	5. Miscellaneous
a. alfalfa	a. rotation pasture	a. spruce-fir	a. wasteland	a. airport
b. barley	b. permanent, nonwooded	b. yellow pine-Douglas fir	b. abandoned farmland	b. school
c. cotton	c. woodland pasture	c. piñon pine-juniper	c. held for non-agricultural development	c. cemetery
d. dates	d. brushland pasture	d. sagebrush (northern desert shrub)		d. park
e. corn		e. creosote (southern desert shrub)		e. mine or quarry
f. fallow		f. greasewood (salt desert shrub)		f. industry
		g. shortgrass (plains grassland)		g. golf course
		h. mesquite grass (desert grassland)		
etc.			etc.	etc.

Fourth Symbol—Quality

1. Cropped Land[4]	2. Pasture Land[4]	3. Forest, Grassland, Shrub[4]	4. Idle Land	5. Miscellaneous
a. good quality	a. good quality	a. merchantable	(no fourth symbol)	
b. moderate	b. moderate	b. potentially merchantable		
c. poor quality	c. poor quality	c. scrub—not merchantable		

1. The RURAL LAND USE mapping key is designed for mapping large areas, i.e., fields, woodlands, etc. For small areas such as farmsteads, buildings, roads, etc., use the RURAL CULTURAL FEATURES KEY.

2. Pasture Land may be defined as follows: *rotation pasture*—presently being used as pasture but will be used as a cropped land at a later date; *permanent pasture*—used as pasture year after year and does not fall into the cropped land rotation system; *woodland pasture*—at least 50 percent is covered by trees, 12 feet or higher; *brushland pasture*—at least 50 percent or more covered by brush, small trees, etc. under 12 feet in height.

3. Applies only to southwestern United States and northern Mexico. This category of land use must be revised to fit the particular area surveyed. Similar subcategories can be developed, however, for all sections of the country and abroad.

4. Quality may be determined subjectively by judging the general condition of the crop or pasture land. If the mapping unit is free of weeds and bare areas, and the crop is uniform in color and height, it may be judged good quality. Quality may also be determined on a measurement basis, i.e., corn yields of 100 bu. per acre—good; 70–100 bu. per acre—moderate; under 70—poor; carrying capacity of pastures, etc.

Rural Cultural Features Key

First Symbol—General Functional Use

1. Farmsteads 2. Nonfarm Country Houses 3. Resort Dwellings 4. Communal Buildings
5. Commercial Buildings 6. Manufactural Structures 7. Roads
8. Irrigation or Drainage Ditches 9. Agricultural Structures

Second Symbol—Specific Functional Use

1. Farmsteads
a. beef cattle
b. citrus
c. dairy
d. poultry
e. cash grain
f. truck
etc.

2. Nonfarm Houses
a. large (7 or more rooms)
b. medium (4 to 6 rooms)
c. small (less than 4 rooms)

3. Resort Dwellings
a. year-round
b. seasonal

4. Communal Buildings
a. school
b. church
c. clubhouse
etc.

5. Commercial Buildings
a. general store
b. food store
c. clothing store
d. hardware store
e. service station
f. tavern
g. drive-in theater
etc.

6. Manufactural Structures
a. gravel pit buildings and plant
b. quarry and buildings
c. cheese factory
d. butter factory
e. cannery
etc.

7. Roads
a. paved (two lane)
b. gravel
c. improved dirt
d. unimproved
e. abandoned
f. farm lane
g. paved (four lane)

8. Irrigation or Drainage Ditches
a. less than 5 ft. wide
b. 5 ft.-10 ft. wide
c. 10 ft.-20 ft. wide
d. 20 ft. or more

9. Agricultural Structures
a. loading pen
b. cotton gin
c. granary
d. cattle vet
e. creamery
etc.

Third Symbol—Age

1. Since 1980 2. 1960–1980 3. 1920–1960 4. Prior to 1920
(Or Other Criteria Suitable to the Specific Area)

Urban Land Use

Second Symbol—General Land Use

1. Residential 2. Commercial 3. Industrial 4. Governmental and Public Utilities 5. Institutional 6. Vacant

Third Symbol—Specific Land Use

1. Residential
a. single family
b. two family
c. multifamily or apartments

2. Commercial
a. appliance store
b. bank, savings and loan
c. cafeteria, restaurant
d. department store
e. drug store
f. furniture store
g. grocery
h. hardware
i. tavern, bar
j. jewelry, gifts
k. beauty salon
l. men's clothing
m. bakery
etc.

3. Industrial
a. food and kindred products
b. tobacco manufacturers
c. textile mill products
d. apparel and clothing
e. lumber and wood products (except furniture)
f. furniture and fixtures
g. paper and allied products
h. printing, publishing, and allied industries
i. chemicals and allied products
j. products of petroleum and coal
etc.

4. Governmental and Public Utilities
a. courthouse
b. fire station
c. police station
d. powerhouse or substation
e. parks or recreation
f. waterworks
g. post office
etc.

5. Institutional
a. school
b. church
c. hospital
etc.

Fourth Symbol—General Condition

1. Well Kept 2. Moderate Condition 3. Deteriorating
(Objective Evaluation Procedure may be Defined Compatible to Area Mapped.)

Fifth Symbol—Age

1. Since 1960 2. 1960–1980 3. 1920–1960 4. Prior to 1920
(Or Other Criteria Suitable to the Specific Area)

The researcher should also be familiar with two standardized land use classification systems that have been designed to be applicable in a variety of environments—*The Standard Land Use Coding Manual*[1] and *A Land Use and Land Cover Classification System for Use with Remote Sensor Data*.[2] The first employs four levels of categories and the second uses three levels of categories. Both have nine generalized first level categories. These classifications should be studied for possible use before other systems are developed. The basic elements of each of these classifications are illustrated below.

First Level Categories

1. Residential
2. Manufacturing (9 second level categories included)
3. Manufacturing (6 second level categories included)
4. Transportation, communications, and utilities
5. Trade
6. Services
7. Cultural, entertainment, and recreation
8. Resource production and extraction
9. Undeveloped land and water areas

Level I
6. Services

Level II
65. Professional services

Level III
651. Medical and other health services

Level IV
6511. Physicians' services
6512. Dental services
6513. Hospital services
6514. Medical laboratory services
6515. Dental laboratory services
6516. Sanitariums, convalescent, and rest home services
6517. Medical clinics—outpatient services
6519. Other medical and health services

First Level Categories2

1. Urban or built-up Land
2. Agricultural land
3. Rangeland
4. Forest Land
5. Water
6. Wetland
7. Barren land
8. Tundra
9. Perennial snow or ice

Level I	Level II	Level III
1. Urban or built-up	11. Residential	111. Single-family Units
		112. Multi-family Units
		113. Group quarters
		114. Residential hotels
		115. Mobile home parks
		116. Transient lodgings
		117. Other

1. *Standard Land Use Coding Manual.* Federal Highway Administration, Bureau of Public Roads. Washington: U.S. Government Printing Office, 1969.
2. Anderson, J. R., Hardy, E. E., Roach, J. T., and Witmer, R. E. *A Land Use and Land Cover Classification System for Use with Remote Sensor Data.* Geological Survey Professional Paper 964, Washington: U.S. Government Printing Office, 1976.

Sample Interview for Geographic Field Study

I f the researcher is planning to collect data by using interviewing techniques, the sample interview sheet for agricultural studies will be of value. If the information is to be obtained indirectly by mail or other mechanisms, the questionnaire must be revised (see page 136–137). Similar interview sheets may be constructed for field studies in other types of geographic studies.

Farm Data Sheet

1. *General*
   a) Name of operator _____
   b) Status (owner, manager, tenant, etc.) _____
   c) How long has he been farming? _____
   d) When did he acquire present farm? _____
   e) Original home area _____
   f) Does he work other farms? _____
   g) Is he employed in work other than farming? _____
       If so, where? _____
       Does he rent, own, sharecrop, etc.? _____

If so, what percentage of his total income is derived from outside sources? _____

_____

_____

_____

2. *Farmstead Population*

Family			*Hired Help*		
Men	Women	Children (under 18)	Men	Women	Children (under 18)
___	___	___	___	___	___

Number full-time _____
Number seasonal _____

3. *Farm, Specific*
    a) Size of farm (acres) _____
    b) Type of farm (dairy, citrus, etc.) _____
    c) Acreage of tilled land (average yield per acre in parenthesis)
       Corn (___) Citrus (___) Cotton (___) Potatoes (___)
       Hay (___) Barley (___) etc. (___) etc. (___) etc. (___)
    d) Acreage of pasture _____

	Permanent		
Rotation	Nonwooded	Wooded	Brush
_____	_____	_____	_____

    e) Other acreage _____

Timber (4 in. plus)	Brush	Waste	Fallow	Idle land (soil bank)	Idle land (other)

    f) Livestock

	Dairy cattle	Beef cattle	Swine	Sheep, Poultry, etc.
Breed	_____	Breed _____	_____	_____
Milkers	_____	Raise own? ___	_____	
Heifers	_____	If brought in,		
Calves	_____	from where? _____		
Bulls	_____			
Total number	_____	_____	_____	_____

How long has dairying been carried on? ____ _____ ____ _____
What was produced before? _____ _____ _____ _____

g) Farming methods
   Rotation system?
   Strip farming? _____ If so, what fields? _____
   Contour plow? _____ If so, what fields? _____
   Irrigation? _____ If so, what fields? _____
   Type of fertilizer used
   How much applied _____ Frequency _____

h) Income (principal commodities sold)

Commodity	% of Total Income	Where Sold
_____ · · ·	_____ · · ·	_____
_____ · · ·	_____ · · ·	_____
_____ · · ·	_____ · · ·	_____
_____ · · ·	_____ · · ·	_____

4. *Trading and Social Centers*

   School _____ Food and drugs _____
   Church _____ Hardware _____
   Movies _____ Clothing _____
   Auto service _____ Farm equipment _____
   Banking center _____ Other _____

_____
_____
_____

5. *Crop hazards and general problems of the area:*

6. *Other pertinent data:*

Sample Checklist for Geographic Field Study

.

If the researcher is planning to make an analysis of the manufacturing geography of an area, the checklist below of pertinent locational and site factors will be of value. The checklist may be abbreviated or otherwise modified without difficulty. Similar checklists may be constructed for field studies in agricultural geography, settlement, trade areas, recreational geography, etc.

General Information

Plant Structure

Products Manufactured—Type of finished product(s) presently being manufactured (if more than one, percent of total value of each); products manufactured in the past, near future (predicted mix of products and rationale).

Plant Organization—Location of branch plants, home office; policy-making structure; relationships to subsidiary plants, branches, and other types of manufacturing plants, etc.

Plant Size and Location—Total acreage, percentage used for buildings, parking, other uses, idle, etc. Location and relationship to transportation and adjacent land use.

Locational Factors

Resource Base

Raw Materials—Location, availability, and price of pertinent agricultural crops, livestock products, forestry commodities, fish and marine life, minerals, other manufactured goods.

Energy—Location, availability, and price of pertinent energy sources such as coal, petroleum, gas, water power. Type of power service, reliability of service, adequacy of supply, rates, discounts, etc.

Water Supplies—Availability of surface and groundwater; quality (mineral and bacteria content) for use as cleaning agent, ingredient in finished product, cooling processes, etc.

Climate—Annual, monthly, maximum-minimum temperature conditions; average degree days; annual rainfall fluctuations; frequency of fog; humidity conditions. General climatic conditions and relationships to living conditions, recreation, etc.

Land Forms and Space—Physiographic restrictions or controls, drainage, and general soil and bedrock characteristics.

Economic Factors

Markets—Location, size, and accessibility of markets.

Transportation—Frequency of service, freight rates, time in transit, terminal facilities, incidental costs of rail, truck, intercoastal or coastwise transportation, air cargo, and local services.

Capital—Availability, magnitude and mobility of capital, fluctuations, time factor, etc.; banking conditions.

Real Estate Value, Taxes, and Service Facilities—Tax rate of real estate and personal property, assessments, special taxes, license fees, exemptions, contemplated future changes in tax base, etc.; service facilities, such as sewers, garbage disposal, police and fire protection, streets and highways, hospital facilities, judiciary, etc.

Cost of Living—Rent, food, clothing, necessities, luxuries, residential rates, etc. on per capita basis.

Labor—Total employment; supply of suitable labor available; elements of labor unrest; past history of labor disturbances; prevailing wage scale; maximum, average, and minimum of hour shifts; labor turnover; characteristics of labor; efficiency of labor; seasonal variation; training facilities; housing; etc.

Social Factors

Type of Culture—Educational level and mores of community; desires to progress; capacity for accepting new ways, etc.
Organization—Organizational and cooperative aspects of community.
Civic Pride—Types and effectiveness of service and fraternal organizations; conditions of schools, churches, libraries, recreational facilities, newspapers, hotels, hospitals, and public buildings.
Technological Skills—Educational level, skills, and general intelligence of community population.

Governmental Factors

National Level—Taxes, tariffs, subsidies, etc.
Local Level—Taxes, existence and mobility of local capital, existence and type of zoning code, effectiveness, and objectives of local and regional planning commissions.

Site Factors

Topographic and Spatial Elevation Characteristics: Size and shape of plot; precise and configuration of land; local drainage conditions; textural and compaction characteristics of the soil; exact nature of the bedrock; water table depth; frost depth; etc.
Service Facilities: Type and cost of sewer and water facilities; type of fire and police protection; garbage disposal conditions; easements; maintenance of access roads, etc.
Transportational Facilities: Rail, highway, and waterway facilities; proximity of piers, interchanges, sidings, terminals, etc.
Water Supplies: Precise volume and nature of potential surface and groundwater supplies.
Land Value and Taxes: Township, county, city tax rate; real estate value and land evaluation trends.
Zoning Characteristics: Zoning restrictions, building codes, contemplated changes in zoning codes; future role of area in overall comprehensive plan, etc.
Proximity to Labor Pool: Commuting conditions, transit time; orientation aspects to local transit lines and residential areas.
Area Trends: Future growth characteristics of local surrounding areas; future restrictions on noise, odors, smoke, etc.

Quantitative Aids

.

The quantitative aids illustrated in this appendix are presented in order that the beginning research student may have available a few of the more common statistical tools used in geography. The aids presented here are not intended as a comprehensive listing, nor is it implied that these tools are suitable for all research projects. The researcher must select the best tool for the type of problem being considered. For additional quantitative aids used in geographic research, consult the bibliographic sources in Appendix A.

I. Correlation-Regression Analysis

Tetrachoric Correlation

TABLE F.1

Coefficients of Tetrachoric Correlations

%	r	%	r	%	r
50	1.00	32	.43	18	-.43
45	.95	31	.37	17	-.49
44	.93	30	.31	16	-.55
43	.91	29	.25	15	-.60
42	.88	28	.19	14	-.65
41	.85	27	.13	13	-.69
40	.81	26	.07	12	-.73
39	.77	25	.00	11	-.77
38	.73	24	-.07	10	-.81
37	.69	23	-.13	9	-.85
36	.65	22	-.19	8	-.88
35	.60	21	-.25	7	-.91
34	.55	20	-.31	6	-.93
33	.49	19	-.37	0	-1.00

From *Short-cut Statistics for Teacher-made Tests* by Paul B. Diederich, © 1964, p. 34, Educational Testing Service, Princeton, New Jersey. Reprinted by permission of the publisher.

Read the coefficient of correlation as the number opposite the percentage of both variables above the median.

Pearson Product Moment Coefficient of Correlation

This is a reliable and commonly used quantitative tool in all types of geographic research. It is included in the appendix rather than in the text so that it may receive individual consideration, and so that it may appear in conjunction with the form for computing correlation-regression analysis which follows it. This coefficient is obtained by first constructing a table of values for the variables being considered. The two variables, X and Y, are then squared and also multiplied in order to obtain the sum for each column to be used in the formuli. The procedure is as follows:

Step 1. Prepare a table of values as shown for the five (N = 5) observations below:

Obs.	X	Y	X^2	Y^2	XY
1	2	4	4	16	8
2	4	8	16	64	32
3	6	12	36	144	72
4	8	16	64	256	128
5	10	20	100	400	200
Sum	30	60	220	880	440

N = 5

Step 2. Compute a figure, called E, as follows:

$$E = \Sigma XY - \frac{(\Sigma X)\ (\Sigma Y)}{N}$$

Note: Σ represents the "sum of" the number it precedes.

In this example, E is thus computed:

$$E = 440 - \frac{(30)\ (60)}{5} \text{ or } 440 - \frac{1800}{5} \text{ or } 440 - 360$$

E = 80

Step 3. Compute two statistics conveniently called F and G. They are computed in the same manner. The formuli are:

$$F = \Sigma X^2 - \frac{(\Sigma X)^2}{N}$$

$$G = \Sigma Y^2 - \frac{(\Sigma Y)^2}{N}$$

F and G are computed as follows:

$$F = 220 - \frac{(30)\ (30)}{5} \text{ or } 220 - \frac{900}{5} \text{ or } 220-180$$

F = 40

$$G = 880 - \frac{(60)\ (60)}{5} \text{ or } 880 - \frac{3600}{5} \text{ or } 880-720$$

G = 160

Step 4. Determine the correlation coefficient by the formula

$$r = \frac{E}{\sqrt{(F)\ (G)}}$$

$$r = \frac{80}{\sqrt{(40)\ (160)}} \text{ or } \frac{80}{\sqrt{6400}} \text{ or } \frac{80}{80}$$

$r = 1.00$

Step 5. Square r to obtain the coefficient of determination, if desired.

$$r^2 = \text{coefficient of determination}$$
$$r^2 = 1^2$$
$$r^2 = 1$$

Regression Analysis

The regression analysis results in a statement of how much the variation of the measurement on the vertical axis (Y) is related to variation of the measurement of the variable (X) on the horizontal axis. The formula for the regression line (Yc) is Yc = a + bX, where a is the elevation of the line at Y axis, and b is the slope of the line.

Using the same table of values and symbols as in the preceding example for the computation of r, the regression line (b) is determined thus:

$$b = \frac{E}{F} \text{ or } b = \frac{80}{40}$$

$b = 2$

The value of Y intercept (a) in the regression equation is determined by the following formula:

$$a = \frac{\Sigma Y - b\Sigma X}{N} \text{ or } \frac{60 - (2)(30)}{5} \text{ or } \frac{60-60}{5} \text{ or } \frac{0}{5}$$

$a = 0$

Consequently, the Y computed by the regression analysis is Yc = 0 + 2X.

Form for Simple Linear Regression and Correlation Analysis (Pearsonian)*

Supply the answer for: Source:
1. Number (N) of observations _____ Table _____
2. Sum (Σ) of columns Table _____
   ΣX _____
   ΣY _____
   ΣX^2 _____
   ΣY^2 _____
   ΣXY _____

---

*For rank order data the regression line (Yc) is computed differently in that the b value is always the same as the Spearman's r and the Y intercept (a) = (N+1) (1- r) ÷ 2.

3. E = _____ $\Sigma XY - \dfrac{(\Sigma X)\,(\Sigma Y)}{N}$

4. F = _____ $\Sigma X^2 - \dfrac{(\Sigma X)^2}{N}$

5. G = _____ $\Sigma Y^2 - \dfrac{(\Sigma Y)^2}{N}$

6. r = _____ $\dfrac{E}{\sqrt{(F)\,(G)}}$

7. r^2 = _____ $(r)\quad(r)$

8. b = _____ $\dfrac{E}{F}$

9. a = _____ $\dfrac{\Sigma Y - b\,(\Sigma X)}{N}$

10. Yc = __+__ (X) $a + b$

II. Locational Quotient

The locational quotient (LQ) of a place is a measure of the extent to which a subregion has a proportionate share of any particular phenomena found in a larger area. It is a ratio of a ratio and is expressed as:

$$LQ = \dfrac{\text{the ratio of phenomena X in a subdivision}}{\text{the ratio of the same phenomena in the larger division}}$$

For example, if a state had 25% of its total population employed, but a city within the state had 50% of its population employed, the LQ of the city would be 2, that is,

$$LQ = \dfrac{50\%}{25\%} \text{ or } 2$$

III. Table of Random Numbers

The table of random numbers is used to select a sample from any universe that has been or can be numbered. To select the sample, use the same number of digits from the column in the table as is in the total number of the universe. That is, if the total of the universe is 8,000, four digits, use all four numbers in the column. The starting place in making the sample may be determined at random, or it may be the beginning of any group in the table. After the first number, continue selecting all numbers from left to right, as in reading. If a number is larger than the total of the universe, ignore it and proceed to the next number until the desired sample is obtained. For example,

TABLE F.2

Significant Values of r, at .05 and .01, for Various Degrees of Freedom

1	.997 1.000	24	.388 .496
2	.950 .990	25	.381 .487
3	.878 .959	26	.374 .478
4	.811 .917	27	.367 .470
5	.754 .874	28	.361 .463
6	.707 .834	29	.355 .456
7	.666 .798	30	.349 .449
8	.632 .765	35	.325 .418
9	.602 .735	40	.304 .393
10	.576 .708	45	.288 .372
11	.553 .684	50	.273 .354
12	.532 .661	60	.250 .325
13	.514 .641	70	.323 .302
14	.497 .623	80	.217 .283
15	.482 .606	90	.205 .267
16	.468 .590	100	.195 .254
17	.456 .575	125	.174 .228
18	.444 .561	150	.159 .208
19	.433 .549	200	.138 .181
20	.423 .537	300	.113 .148
21	.413 .526	400	.098 .128
22	.404 .515	500	.088 .115
23	.396 .505	1000	.062 .081
		x	

Source: William Hays. *Basic Statistics.* Brooks Cole, Belmonte, California, 1967. p. 114.

if from a universe of 8,000, a sample of 80 is desired, and the starting point is from group 1, line 11, column 1, the sample would be as follows: 0709, 2523, skip, 6271, 2607, and so forth, until 80 sample numbers are selected. The third number in the table, 9224, was not considered since it was larger than the universe of 8,000. It was not counted as one number in the sample of 80 numbers.

TABLE F.3

Table of Random Numbers (8,000 Numbers)

First Thousand

	1–4	5–8	9–12	13–16	17–20	21–24	25–28	29–32	33–36	37–40
1	23 15	75 48	59 01	83 72	59 93	76 24	97 08	86 95	23 03	67 44
2	05 54	55 50	43 10	53 74	35 08	90 61	18 37	44 10	96 22	13 43
3	14 87	16 03	50 32	40 43	62 23	50 05	10 03	22 11	54 38	08 34
4	38 97	67 49	51 94	05 17	58 53	78 80	59 01	94 32	42 87	16 95
5	97 31	26 17	18 99	75 53	08 70	94 25	12 58	41 54	88 21	05 13
6	11 74	26 93	81 44	33 93	08 72	32 79	73 31	18 22	64 70	68 50
7	43 36	12 88	59 11	01 64	56 23	93 00	90 04	99 43	64 07	40 36
8	93 80	62 04	78 38	26 80	44 91	55 75	11 89	32 58	47 55	25 71
9	49 54	01 31	81 08	42 98	41 87	69 53	82 96	61 77	73 80	95 27
10	36 76	87 26	33 37	94 82	15 69	41 95	96 86	70 45	27 48	38 80
11	07 09	25 23	92 24	62 71	26 07	06 55	84 53	44 67	33 84	53 20
12	43 31	00 10	81 44	86 38	03 07	52 55	51 61	48 89	74 29	46 47
13	61 57	00 63	60 06	17 36	37 75	63 14	89 51	23 35	01 74	69 93
14	31 35	28 37	99 10	77 91	89 41	31 57	97 64	48 62	58 48	69 19
15	57 04	88 65	26 27	79 59	36 82	90 52	95 65	46 35	06 53	22 54
16	09 24	34 42	00 68	72 10	71 37	30 72	97 57	56 09	29 82	76 50
17	97 95	53 50	18 40	89 48	83 29	52 23	08 25	21 22	53 26	15 87
18	93 73	25 95	70 43	78 19	88 85	56 67	16 68	26 95	99 64	45 69
19	72 62	11 12	25 00	92 26	82 64	35 66	65 94	34 71	68 75	18 67
20	61 02	07 44	18 45	37 12	07 94	95 91	73 78	66 99	53 61	93 78
21	97 83	98 54	74 33	05 59	17 18	45 47	35 41	44 22	03 42	30 00
22	89 16	09 71	92 22	23 29	06 37	35 05	54 54	89 88	43 81	63 61
23	25 96	68 82	20 62	87 17	92 65	02 82	35 28	62 84	91 95	48 83
24	81 44	33 17	19 05	04 95	48 06	74 69	00 75	67 65	01 71	65 45
25	11 32	25 49	31 42	36 23	43 86	08 62	49 76	67 42	24 52	32 45

Second Thousand

	1–4	5–8	9–12	13–16	17–20	21–24	25–28	29–32	33–36	37–40
1	64 75	58 38	85 84	12 22	59 20	17 69	61 56	55 95	04 59	59 47
2	10 30	25 22	89 77	43 63	44 30	38 11	24 90	67 07	34 82	33 28
3	71 01	79 84	95 51	30 85	03 74	66 59	10 28	87 53	76 56	91 49
4	60 01	25 56	05 88	41 03	48 79	79 65	59 01	69 78	80 00	36 66
5	37 33	09 46	56 49	16 14	28 02	48 27	45 47	55 44	55 36	50 90
6	47 86	98 70	01 31	59 11	22 73	60 62	61 28	22 34	69 16	12 12
7	38 04	04 27	37 64	16 78	95 78	39 32	34 93	24 88	43 43	87 06
8	73 50	83 09	08 83	05 48	00 78	36 66	93 02	95 56	46 04	53 36
9	32 62	34 64	74 84	06 10	43 24	20 62	83 73	19 32	35 64	39 69
10	97 59	19 95	49 36	63 03	51 06	62 06	99 29	75 95	32 05	77 34
11	74 01	23 19	55 59	79 09	69 82	66 22	42 40	15 96	74 90	75 89
12	56 75	42 64	57 13	35 10	50 14	90 96	63 36	74 69	09 63	34 88
13	49 80	04 99	08 54	83 12	19 98	08 52	82 63	72 92	92 36	50 26
14	43 58	48 96	47 24	87 85	66 70	00 22	15 01	93 99	59 16	23 77
15	16 65	37 96	64 60	32 57	13 01	35 74	28 36	36 73	05 88	72 29
16	48 50	26 90	55 65	32 25	87 48	31 44	68 02	37 31	25 29	63 67
17	96 76	55 46	92 36	31 68	62 30	48 29	63 83	52 23	81 66	40 94
18	38 92	36 15	50 80	35 78	17 84	23 44	41 24	63 33	99 22	81 28
19	77 95	88 16	94 25	22 50	55 87	51 07	30 10	70 60	21 86	19 61
20	17 92	82 80	65 25	58 60	87 71	02 64	18 50	64 65	79 64	81 70
21	94 03	68 59	78 02	31 80	44 99	41 05	41 05	31 87	43 12	15 96
22	47 46	06 04	79 56	23 04	84 17	14 37	28 51	67 27	55 80	03 68
23	47 85	65 60	88 51	99 28	24 39	40 64	41 71	70 13	46 31	82 88
24	57 61	63 46	53 92	29 86	20 18	10 37	57 65	15 62	98 69	07 56
25	08 30	09 27	04 66	75 26	66 10	57 18	87 91	07 54	22 22	20 13

From *Tables of Random Sampling Numbers* by M. G. Kendall and B. B. Smith, (London: Cambridge University Press, © 1939), pp. 2—5. Reprinted by permission of the publisher.

TABLE F.3 Continued

	1–4	5–8	9–12	13–16	17–20	21–24	25–28	29–32	33–36	37–40
					Third Thousand					
1	89 22	10 23	62 65	78 77	47 33	51 27	23 02	13 92	44 13	96 51
2	04 00	59 98	18 63	91 82	90 32	94 01	24 23	63 01	26 11	06 50
3	98 54	63 80	66 50	85 67	50 45	40 64	52 28	41 53	25 44	41 25
4	41 71	98 44	01 59	22 60	13 14	54 58	14 03	98 49	98 86	55 79
5	28 73	37 24	89 00	78 52	58 43	24 61	34 97	97 85	56 78	44 71
6	65 21	38 39	27 77	76 20	30 86	80 74	22 43	95 68	47 68	37 92
7	65 55	31 26	78 90	90 69	04 66	43 67	02 62	17 69	90 03	12 05
8	05 66	86 90	80 73	02 98	57 46	58 33	27 82	31 45	98 69	29 98
9	39 30	29 97	18 49	75 77	95 19	27 38	77 63	73 47	26 29	16 12
10	64 59	23 22	54 45	87 92	94 31	38 32	00 59	81 18	06 78	71 37
11	07 51	34 87	92 47	31 48	36 60	68 90	70 53	36 82	57 99	15 82
12	86 59	36 85	01 56	63 89	98 00	82 83	93 51	48 56	54 10	72 32
13	83 73	52 25	99 97	97 78	12 48	36 83	89 95	60 32	41 06	76 14
14	08 59	52 18	26 54	65 50	82 04	87 99	01 70	33 56	25 80	53 84
15	41 27	32 71	49 44	29 36	94 58	16 82	86 39	62 15	86 43	54 31
16	00 47	37 59	08 56	23 81	22 42	72 63	17 63	17 47	25 20	63 47
17	86 13	15 37	89 81	38 30	78 68	89 13	29 61	82 07	00 98	64 32
18	33 84	97 83	59 04	40 20	35 86	03 17	68 86	63 08	01 82	25 46
19	61 87	04 16	57 07	46 80	86 12	98 08	39 73	49 20	77 54	50 91
20	43 89	86 59	23 25	07 88	61 29	78 49	19 76	53 91	50 08	07 86
21	29 93	93 91	23 04	54 84	59 85	60 95	20 66	41 28	72 64	64 73
22	38 50	58 55	55 14	38 85	50 77	18 65	79 48	87 67	83 17	08 19
23	31 82	43 84	31 67	12 52	55 11	72 04	41 15	62 53	27 98	22 68
24	91 43	00 37	67 13	56 11	55 97	06 75	09 25	52 02	39 13	87 53
25	38 63	56 89	70 25	49 89	75 26	96 45	80 38	05 04	11 66	35 14
					Fourth Thousand					
1	02 49	05 41	22 27	94 43	93 64	04 23	07 20	74 11	67 95	40 82
2	11 96	73 64	69 60	62 78	37 01	09 25	33 02	08 01	38 53	74 82
3	48 25	68 34	65 49	69 92	40 79	05 40	33 51	54 39	61 30	31 36
4	27 24	67 30	80 21	48 12	35 36	04 88	18 99	77 49	48 49	30 71
5	32 53	27 72	65 72	43 07	07 22	86 52	91 84	57 92	65 71	00 11
6	66 75	79 89	55 92	37 59	34 31	43 20	45 58	25 45	44 36	92 65
7	11 26	63 45	45 76	50 59	77 46	34 66	82 69	99 26	74 29	75 16
8	17 87	23 91	42 45	56 18	01 46	93 13	74 89	24 64	25 75	92 84
9	62 56	13 03	65 03	40 81	47 54	51 79	80 81	33 61	01 09	77 30
10	62 79	63 07	79 35	49 77	05 01	30 10	50 81	33 00	99 79	19 70
11	75 51	02 17	71 04	33 93	36 60	42 75	76 22	23 87	56 54	84 68
12	87 43	90 16	91 63	51 72	65 90	44 43	70 72	17 98	70 63	90 32
13	97 74	20 26	21 10	74 87	88 03	38 33	76 52	26 92	14 95	90 51
14	98 81	10 60	01 21	57 10	28 75	21 82	88 39	12 85	18 86	16 24
15	51 26	40 18	52 64	60 79	25 53	29 00	42 66	95 78	58 36	29 98
16	40 23	99 33	76 10	41 96	86 10	49 12	00 29	41 80	03 59	93 17
17	26 93	65 91	86 51	66 72	76 45	46 32	94 46	81 94	19 06	66 47
18	88 50	21 17	16 98	29 94	09 74	42 39	46 22	00 69	09 48	16 46
19	63 49	93 80	93 25	59 36	19 95	79 86	78 05	69 01	02 33	83 74
20	36 37	98 12	06 03	31 77	87 10	73 82	83 10	83 60	50 94	40 91
21	93 80	12 23	22 47	47 95	70 17	59 33	43 06	47 43	06 12	66 60
22	29 85	68 71	20 56	31 15	00 53	25 36	58 12	65 22	41 40	24 31
23	97 72	08 79	31 88	26 51	30 50	71 01	71 51	77 06	95 79	29 19
24	85 23	70 91	05 74	60 14	63 77	59 93	81 56	47 34	17 79	27 53
25	75 74	67 52	68 31	72 79	57 73	72 36	48 73	24 36	87 90	68 02

TABLE F.3 Continued

	1–4	5–8	9–12	13–16	17–20	21–24	25–28	29–32	33–36	37–40
Fifth Thousand										
1	29 93	50 69	71 63	17 55	25 79	10 47	88 93	79 61	42 82	13 63
2	15 11	40 71	26 51	89 07	77 87	75 51	01 31	03 42	94 24	81 11
3	03 87	04 32	25 10	58 98	76 29	22 03	99 41	24 38	12 76	50 22
4	79 39	03 91	88 40	75 64	52 69	65 95	92 06	40 14	28 42	29 60
5	30 03	50 69	15 79	19 65	44 28	64 81	95 23	14 48	72 18	15 94
6	29 03	99 98	61 28	75 97	98 02	68 53	13 91	98 38	13 72	43 73
7	78 19	60 81	08 24	10 74	97 77	09 59	94 35	69 84	82 09	49 56
8	15 84	78 54	93 91	44 29	13 51	80 13	07 37	52 21	53 91	09 86
9	36 61	46 22	48 49	19 49	72 09	92 58	79 20	53 41	02 18	00 64
10	40 54	95 48	84 91	46 54	38 62	35 54	14 44	66 88	89 47	41 80
11	40 87	80 89	97 14	28 60	99 82	90 30	87 80	07 51	58 71	66 58
12	10 22	94 92	82 41	17 33	14 68	59 45	51 87	56 08	90 80	66 60
13	15 91	87 67	87 30	62 42	59 28	44 12	42 50	88 31	13 77	16 14
14	13 40	31 87	96 49	90 99	44 04	64 97	94 14	62 18	15 59	83 35
15	66 52	39 45	96 74	90 89	02 71	10 00	99 86	48 17	64 06	89 09
16	91 66	53 64	69 68	34 31	78 70	25 97	50 46	62 21	27 25	06 20
17	67 41	58 75	15 08	20 77	37 29	73 20	15 75	93 96	91 76	96 99
18	76 52	79 69	96 23	72 43	34 48	63 39	23 23	94 60	88 79	06 17
19	19 81	54 77	89 74	34 81	71 47	10 95	43 43	55 81	19 45	44 07
20	25 59	25 35	87 76	38 47	25 75	84 34	76 89	18 05	73 95	72 22
21	55 90	24 55	36 63	64 63	16 09	95 99	98 28	87 40	66 66	66 92
22	02 47	05 83	76 79	79 42	24 82	42 42	39 61	62 47	49 11	72 64
23	18 63	05 32	63 13	31 99	76 19	35 85	91 23	50 14	63 28	86 59
24	89 67	33 82	30 16	06 39	20 07	59 50	33 84	02 76	45 03	33 33
25	62 98	66 73	64 06	59 51	74 27	84 62	31 45	65 82	86 05	73 00
Sixth Thousand										
	1–4	5–8	9–12	13–16	17–20	21–24	25–28	29–32	33–36	37–40
1	27 50	13 05	46 34	63 85	87 60	35 55	05 67	88 15	47 00	50 92
2	02 31	57 57	62 98	41 09	66 01	69 88	92 83	35 70	76 59	02 58
3	37 43	12 83	66 39	77 33	63 26	53 99	48 65	23 06	94 29	53 04
4	83 56	65 54	19 33	35 42	92 12	37 14	70 75	18 58	98 57	12 52
5	06 81	56 27	49 32	12 42	92 42	05 96	82 94	70 25	45 49	18 16
6	39 15	03 60	15 56	73 16	48 74	50 27	43 42	58 36	73 16	39 90
7	84 45	71 93	10 27	15 83	84 20	57 42	41 28	42 06	15 90	70 47
8	82 47	05 77	06 89	47 13	92 85	60 12	32 89	25 22	42 38	87 37
9	98 04	06 70	24 21	69 02	65 42	55 33	11 95	72 35	73 23	57 26
10	18 33	49 04	14 33	48 50	15 64	58 26	14 91	46 02	72 13	48 62
11	33 92	19 93	38 27	43 40	27 72	79 74	86 57	41 83	58 71	56 99
12	48 66	74 30	44 81	06 80	29 09	50 31	69 61	24 64	28 89	97 79
13	85 85	07 54	21 50	31 80	10 19	56 65	82 52	26 58	55 12	26 34
14	08 27	08 08	35 87	96 57	33 12	01 77	52 76	09 89	71 12	17 69
15	59 61	22 14	26 09	96 75	17 94	51 08	41 91	45 94	80 48	59 92
16	17 45	77 79	31 66	36 54	92 85	65 60	53 98	63 50	11 20	96 63
17	11 26	37 08	07 71	95 95	39 75	92 48	99 78	23 33	19 56	06 67
18	48 08	13 98	16 52	41 15	73 96	32 55	03 12	38 30	88 77	17 03
19	76 27	72 22	99 61	72 15	00 25	21 54	47 79	18 41	58 50	57 66
20	98 89	22 25	72 92	53 55	07 98	66 71	53 29	61 71	56 96	41 78
21	88 69	61 63	01 67	61 88	58 79	35 65	08 45	63 38	69 86	79 47
22	12 58	13 75	80 98	01 35	91 16	18 36	90 54	99 17	68 36	85 06
23	08 86	96 36	14 09	43 85	51 20	65 18	06 40	52 17	48 10	68 97
24	33 81	05 51	32 48	60 12	32 44	08 12	89 00	98 82	79 17	97 22
25	05 15	99 28	87 15	07 08	66 92	53 81	69 42	02 27	65 33	57 69

TABLE F.3 Continued

Seventh Thousand

	1–4	5–8	9–12	13–16	17–20	21–24	25–28	29–32	33–36	37–40
1	80 30	23 64	67 96	21 33	36 90	03 91	69 33	90 13	34 48	02 19
2	61 29	89 61	32 08	12 62	26 08	42 00	31 73	31 30	30 61	34 11
3	23 33	61 01	02 21	11 81	51 32	36 10	23 74	50 31	90 11	73 52
4	94 21	32 92	93 50	72 67	23 20	74 59	30 30	48 66	75 32	27 97
5	87 61	92 69	01 60	28 79	74 76	86 06	39 29	73 85	03 27	50 57
6	37 56	19 18	03 42	86 03	85 74	44 81	86 45	71 16	13 52	35 56
7	64 86	66 31	55 04	88 40	10 30	84 38	06 13	58 83	62 04	63 52
8	22 69	58 45	49 23	09 81	98 84	05 04	75 99	27 70	72 79	32 19
9	23 22	14 22	64 90	10 26	74 23	53 91	27 73	78 19	92 43	68 10
10	42 38	59 64	72 96	46 57	89 67	22 81	94 56	69 84	18 31	06 39
11	17 18	01 34	10 98	37 48	93 86	88 59	69 53	78 86	37 26	85 48
12	39 45	69 53	94 89	58 97	29 33	29 19	50 94	80 57	31 99	38 91
13	43 18	11 42	56 19	48 44	45 02	84 29	01 78	65 77	76 84	88 85
14	59 44	06 45	68 55	16 65	66 13	38 00	95 76	50 67	67 65	18 83
15	01 50	34 32	38 00	37 57	47 82	66 59	19 50	87 14	35 59	79 47
16	79 14	60 35	47 95	90 71	31 03	85 37	38 70	34 16	64 55	66 49
17	01 56	63 68	80 26	14 97	23 88	59 22	82 39	70 83	48 34	46 48
18	25 76	18 71	29 25	15 51	92 96	01 01	28 18	03 35	11 10	27 84
19	23 52	10 83	45 06	49 85	35 45	84 08	81 13	52 57	21 23	67 02
20	91 64	08 64	25 74	16 10	97 31	10 27	24 48	89 06	42 81	29 10
21	80 86	07 27	26 70	08 65	85 20	31 23	28 99	39 63	32 03	71 91
22	31 71	37 60	95 60	94 95	54 45	27 97	03 67	30 54	86 04	12 41
23	05 83	50 36	09 04	39 15	66 55	80 36	39 71	24 10	62 22	21 53
24	98 70	02 90	30 63	62 59	26 04	97 20	00 91	28 80	40 23	09 91
25	82 79	35 45	64 53	93 24	86 55	48 72	18 57	05 79	20 09	31 46

Eighth Thousand

	1–4	5–8	9–12	13–16	17–20	21–24	25–28	29–32	33–36	37–40
1	37 52	49 55	40 65	27 61	08 59	91 23	26 18	95 04	98 20	99 52
2	48 16	69 65	69 02	08 83	08 83	68 37	00 96	13 59	12 16	17 93
3	50 43	06 59	56 53	30 61	40 21	29 06	49 60	90 38	21 43	19 25
4	89 13	62 79	45 73	71 72	77 11	28 80	72 35	75 77	24 72	98 43
5	63 29	90 61	86 39	07 38	38 85	77 06	10 23	30 84	07 95	30 76
6	71 68	93 94	08 72	36 27	85 89	40 59	83 37	93 85	73 97	84 05
7	05 06	96 63	58 24	05 95	56 64	77 53	85 64	15 95	93 91	59 03
8	03 35	58 95	46 44	25 70	31 66	01 05	44 44	62 91	36 31	45 04
9	13 04	57 67	74 77	53 35	93 51	82 83	27 38	63 16	04 48	75 23
10	49 96	43 94	56 04	02 79	55 78	01 44	75 26	85 54	01 81	32 82
11	24 36	24 08	44 77	57 07	54 41	04 56	09 44	30 58	25 45	37 56
12	55 19	97 20	01 11	47 45	79 79	06 72	12 81	86 97	54 09	06 53
13	02 28	54 60	28 35	32 94	36 74	51 63	96 90	04 13	30 43	10 14
14	90 50	13 78	22 20	37 56	97 95	49 95	91 15	52 73	12 93	78 94
15	33 71	32 43	29 58	47 38	39 96	67 51	64 47	49 91	64 58	93 07
16	70 58	28 49	54 32	97 70	27 81	64 69	71 52	02 56	61 37	04 58
17	09 68	96 10	57 78	85 00	89 81	98 30	19 40	76 28	62 99	99 83
18	19 36	60 85	35 04	12 87	83 88	66 54	32 00	30 20	05 30	42 63
19	04 75	44 49	64 26	51 46	80 50	53 91	00 55	07 01	68 66	08 29
20	79 83	32 39	46 77	56 83	42 21	60 03	14 47	07 01	66 85	49 22
21	80 99	42 43	08 58	54 41	98 05	54 39	34 42	97 47	38 35	59 40
22	48 83	64 99	86 94	48 78	79 20	62 23	56 45	92 65	56 36	83 02
23	28 45	35 85	22 20	13 01	73 96	70 05	84 50	68 59	96 58	16 63
24	52 07	63 15	82 30	66 23	14 26	66 61	17 80	41 97	40 27	24 80
25	39 14	52 18	35 87	48 55	48 81	03 11	26 99	03 80	08 86	50 42

Measurement Scales

.

Measure Scales

G.1 AREA (SQUARE) MEASURE

Square/Square Feet	Meters	Acres/ Hectares		Square/Square Miles	Kilometers
1	.0929	1	.40469	1	2.590
2	.1858	2	.80937	2	5.180
3	.2787	3	1.2141	3	7.770
4	.3716	4	1.6188	4	10.36
5	.4645	5	2.0234	5	12.95
6	.5574	6	2.4281	6	15.54
7	.6503	7	2.8328	7	18.13
8	.7432	8	3.2375	8	20.72
1 sq. yd. — 9	.8361	9	3.6422	9	23.31
10	.9290	10	4.0469	10	25.90
11	1.022	11	4.4516	11	28.49
12	1.115	12	4.8562	12	31.08
13	1.208	13	5.2609	13	33.67
14	1.301	14	5.6656	14	36.26
15	1.394	15	6.0703	15	38.85
16	1.486	16	6.4750	16	41.44
17	1.579	17	6.8797	17	44.03
2 sq. yds — 18	1.672	18	7.2844	18	46.62

1 sq. ft. = .0929 sq. m.	1 acre = .404687 hectare	1 sq. mi. = 2.590 sq. km.
1 sq. yd. = .8361 sq. m.	1 hectare = 2.4710 acres	1 sq. km. = .3861 sq. mi.
1 sq. m. = 10.764 sq. ft.	1 hectare = 10,000 sq. m.	1 sq. km. = 247.104 acres

G.2 LINEAR MEASURE

Inches/Centimeters		Yards/Meters		Miles/Kilometers	
1	2.54	1	.914	5	8.047
2	5.08	2	1.82	10	16.09
3	7.62	3	2.74	15	24.14
4	10.16	4	3.65	20	32.18
5	12.70	5	4.57	25	40.23
6	16.24	6	5.48	30	48.27
7	17.78	7	6.39	35	56.32
8	20.32	8	7.31	40	64.36
9	22.86	9	8.22	45	72.42
10	25.40	10	9.14	50	80.45
11	27.94	11	10.05	55	88.51
12	30.48	12	10.97	60	96.54
13	33.02	13	11.88	65	104.61
14	35.56	14	12.80	70	112.63
15	38.10	15	13.71	75	120.70
16	40.64	16	14.62	80	128.72
17	43.18	17	15.54	85	136.79
18	45.72	18	16.45	90	144.81

1 in. = 2.54 cm.	1 yd. = 0.914 m.	1 mi. = 1.6093 km.
1 cm. = 0.3937 in.	1 m. = 1.093 yd.	1 km. = 0.62137 mi.

G.3 TEMPERATURE

Temperature in Celsius = 5/9 (temperature in °F −32°)
Temperature in Fahrenheit = 9/5 (temperature in °C) + 32°
Temperature in Kelvin or Absolute = temperature in °C −273.15°

Fahrenheit to Celsius (Centigrade)

The vertical column on the left indicates 10°F intervals; 1°F intervals are shown across the top.

Degrees Fahrenheit	Degrees Centigrade									
	0	1	2	3	4	5	6	7	8	9
110	43.3	43.9	44.4	45.0	45.6	46.1	46.7	47.2	47.8	48.3
100	37.8	38.3	38.9	39.4	40.0	40.6	41.1	41.7	42.2	42.8
90	32.2	32.8	33.3	33.9	34.4	35.0	35.6	36.1	36.7	37.2
80	26.7	27.2	27.8	28.3	28.9	29.4	30.0	30.6	31.1	31.7
70	21.1	21.7	22.2	22.8	23.3	23.9	24.4	25.0	25.6	26.1
60	15.6	16.1	16.7	17.2	17.8	18.3	18.9	19.4	20.0	20.6
50	10.0	10.6	11.1	11.7	12.2	12.8	13.3	13.9	14.4	15.0
40	4.4	5.0	5.6	6.1	6.7	7.2	7.8	8.3	8.9	9.4
30	−1.1	−0.6	0.0	0.6	1.1	1.7	2.2	2.8	3.3	3.9
20	−6.7	−6.1	−5.6	−5.0	−4.4	−3.9	−3.3	−2.8	−2.2	−1.7
10	−12.2	−11.7	−11.1	−10.6	−10.0	−9.4	−8.9	−8.3	−7.8	−7.2
+0	−17.8	−17.2	−16.7	−16.1	−15.6	−15.0	−14.4	−13.9	−13.3	−12.8
−0	−17.8	−18.3	−18.9	−19.4	−20.0	−20.6	−21.1	−21.7	−22.2	−22.8
−10	−23.3	−23.9	−24.4	−25.0	−25.6	−26.1	−26.7	−27.2	−27.8	−28.3
−20	−28.9	−29.4	−30.0	−30.6	−31.1	−31.7	−32.2	−32.8	−33.3	−33.9
−30	−34.4	−35.0	−35.6	−36.1	−36.7	−37.2	−37.8	−33.3	−38.9	−39.1
−40	−40.0	−40.6	−41.1	−41.7	−42.2	−42.8	−43.3	−43.9	−44.4	−45.0
−50	−45.6	−46.1	−46.7	−47.2	−47.8	−48.3	−48.9	−49.4	−50.0	−50.6
−60	−51.1	−51.7	−52.2	−52.8	−53.3	−53.9	−54.4	−55.0	−55.6	−56.1

Celsius (Centigrade) to Fahrenheit

The vertical column on the left indicates 10°C intervals; 1°C intervals are shown across the top.

Degrees Centigrade	Degrees Fahrenheit									
	0	1	2	3	4	5	6	7	8	9
40	104.0	105.8	107.6	109.4	111.2	113.0	114.8	116.6	118.4	120.2
30	86.0	87.8	89.6	91.4	93.2	95.0	96.8	98.6	100.4	102.2
20	68.0	69.8	71.6	73.4	75.2	77.0	78.8	80.6	82.4	84.2
10	50.0	51.8	53.6	55.4	57.2	59.0	60.8	62.6	64.4	66.2
+0	32.0	33.8	35.6	37.4	39.2	41.0	42.8	44.6	46.4	48.2
−0	32.0	30.2	28.4	26.6	24.8	23.0	21.2	19.4	17.6	15.8
−10	14.0	12.2	10.4	8.6	6.8	5.0	3.2	1.4	−0.4	−2.2
−20	−4.0	−5.8	−7.6	−9.4	−11.2	−13.0	−14.8	−16.6	−18.4	−20.2
−30	−22.0	−23.8	−25.6	−27.4	−29.2	−31.0	−32.8	−34.6	−36.4	−38.2
−40	−40.0	−41.8	−43.6	−45.4	−47.2	−49.0	−50.8	−52.6	−54.4	−56.2
−50	−58.0	−59.8	−61.6	−63.4	−65.2	−67.0	−68.8	−70.6	−72.4	−74.2

G.4 MASS

1 ounce (av) = 28.35 grams
1 gram = 0.0353 ounce (av)
1 kilogram = 2.2046 pounds (av) = 1000 grams
1 metric ton = 1000 kilograms = 2204.6 pounds (av) = 1.10 short tons

G.5 VOLUME

1 cubic foot = 0.028 cubic meter
1 cubic meter = 1.308 cubic yards = 35.31 cubic feet = 61,024 cubic inches
1 cubic kilometer = 0.2399 cubic mile

G.6 VELOCITY

1 knot (nautical mile) = 1.1516 statute miles per hour = 0.5144 meter per second = 1.85 kilometers per hour
1 meter per second = 3.281 feet per second = 1.942 nautical miles per hour = 2.237 statute miles per hour = 3.6 kilometers per hour

G.7 ENERGY/POWER

1 calorie = 4.186 joules = 3.968×10^{-3} British thermal units
1 joule = 0.738 foot pound = 10^7 ergs
1 British thermal unit = 251.98 calories = 1055 joules = 0.293 watt-hour

1 langley = 1 calorie per square centimeter
1 horse power = 746 watts = 33,000 foot pounds per minute
1 calorie per minute = 251.16 watts
Solar constant = approximately 2 langleys per minute

INDEX